# OPTIMIZATION
# OF SYSTEMS RELIABILITY

INDUSTRIAL ENGINEERING

*A Series of Reference Books and Textbooks*

*Editor*

WILBUR MEIER, JR.
*Head, School of Industrial Engineering*
*Purdue University*
*West Lafayette, Indiana*



Additional Volumes in Preparation

# OPTIMIZATION OF SYSTEMS RELIABILITY

Frank A. Tillman
Ching-Lai Hwang
Way Kuo

Department of Industrial Engineering
Kansas State University
Manhattan, Kansas

MARCEL DEKKER, INC.

New York and Basel

Library of Congress Cataloging in Publication Data

Main entry under title:

Optimization of systems reliability.

   (Industrial engineering ; v. 4)
   Bibliography: p.
   Includes index.
   1.  Reliability (Engineering)  2.  Mathematical
optimization.  I.  Tillman, Frank A.
II. Hwang, Ching Lai.   III.  Kuo, Way.
IV.  Series.
TS173.067        620.00452          80-22660
ISBN 0-8247-6989-9





MARCEL DEKKER, INC.

270 Madison Avenue, New York, New York  10016


Current printing (last digit):

10  9  8  7  6  5  4  3  2  1

PRINTED IN THE UNITED STATES OF AMERICA

This book presents the current methods used to optimize systems reliability problems; it is a comprehensive classification and presentation of problems and techniques useful in solving systems reliability problems.  This book is not only useful to reliability engineers but for people working and teaching in the field of operations research in that it contains a comprehensive collection of techniques.

A state-of-the-art review of the literature is presented in Chapter 2.  The literature is classified according to:  (1) system models, (2) optimization techniques, and (3) various reliability problems.

Each technique has strengths and weaknesses for solving various general linear or nonlinear programming problems and these are illustrated by solving a number of systems reliability problems.  The optimization techniques presented are:

1.  heuristic approaches
2.  dynamic programming
3.  the discrete maximum principle
4.  the sequential unconstrained minimization technique (SUMT)
5.  the generalized reduced gradient method (GRG)
6.  method of Lagrange multipliers and the Kuhn-Tucker conditions
7.  the generalized Lagrangian functions method

8.  geometric programming

9.  integer programming

10. others (a classical approach, a parametric method, linear programming, and separable programming).

In Chapter 13, the usual constrained reliability optimization problem is extended to include the determination of the optimal level of component reliability and the number of redundancies in each stage. This type of problem is a mixed integer nonlinear programming problem. A new approach is suggested to solve this problem which is a combination of the Hooke and Jeeves pattern search and a heuristic technique.

## ACKNOWLEDGMENTS

We thank Mrs. Marie Davis and Mrs. Merla Oppy for their excellent typing, and Mrs. Jean Burnham for her editing. This book is an outgrowth of our research in this field for the past twelve years which has been supported by NASA, the Air Force Office of Scientific Research, and the Office of Naval Research.

Chapter 2 is based on F. A. Tillman, C. L. Hwang, and W. Kuo, "Optimization techniques for systems reliability with redundancy - a review," *IEEE Trans. on Reliability,* Vol. R-26, pp. 148-155 (1977); Section 3.2 is based on J. Sharma and K. V. Venkateswaran, "A direct method for maximizing the system reliability," *IEEE Trans. on Reliability,* Vol. R-20, pp. 256-259 (1971); Section 3.3 is based on K. K. Aggarwal, J. S. Gupta, and K. B. Misra "A new heuristic criterion for solving a redundancy optimization problem," *IEEE Trans. on Reliability,* Vol. R-24, pp 86-87 (1975); Section 3.4 is based on K. B. Misra, "A simple approach for constrained redundancy optimization problem," *IEEE Trans. on Reliability,* Vol. R-21, pp. 30-34 (1972); Section 3.6 is based on Y. Nakagawa and K. Nakashima, "A heuristic method for determining optimal reliability allocation," *IEEE Trans. on Reliability,* Vol. R-26, pp. 156-161 (1977), and W. Kuo, C. L. Hwang and F. A. Tillman, "A note on heuristic methods in optimal system reliability," *IEEE Trans. on Reliability,* Vol. R-27, pp. 320-324 (1978); Chapter 5 is based on F. A. Tillman, C. L. Hwang, L. T. Fan, and S. A. Balbale, "System reliability subject to multiple nonlinear constraints," *IEEE Trans. on Reliability,* Vol. R-17, pp. 153-157 (1968); Chapter 6 is based on F. A. Tillman, C. L. Hwang, L. T. Fan and K. C. Lai, "Optimal reliability of a complex system," *IEEE Trans. on Reliability,* Vol. R-19, pp. 95-100 (1970), and C. L. Hwang, K. C. Lai, F. A. Tillman and L. T. Fan, "Optimization of system reliability by the sequential unconstrained minimization techniques," *IEEE Trans. on Reliability,* Vol. R-24, pp. 133-135 (1975); Parts of Chapter 7 and Chapter 9 are based on C. L. Hwang, F. A. Tillman, and W. Kuo, "Reliability optimization by generalized Lagrangian-function and reduced-gradient methods," *IEEE Trans. on Reliability,* Vol. R-28, pp. 316-319 (1979); Section 11.5 is based on K. N. Hyun, "Reliability optimization by 0-1 programming for a system with several failure modes," *IEEE Trans. on Reliability,* Vol. R-24, pp. 206-210 (1975); Section 12.4 is based on V. Selman and N. T. Grisamore, "Optimum system analysis by linear programming," *1966 Proc. Annual Symposium and Reliability,* pp. 696-703 (1966); Chapter 13 is based on F. A. Tillman, C. L. Hwang, and W. Kuo, "Determining component reliability and redundancy for optimum system reliability," *IEEE Trans. on Reliability,* Vol. R-26, pp. 162-165 (1977).  The foregoing are all reprinted with permission of the Institute of Electrical and Electronics Engineers, Inc.

Section 3.5 is based on I. A. Ushakov, "A heuristic method of optimization of the redundancy of multi-function system," *Engineering Cybernetics,* Vol. 10, No. 4, pp. 612-613 (1972).  This is reprinted with permission of the Scripta Publishing Company.

Section 8.2 is based on H. Everett III, "Generalized Lagrange multiplier method for solving problems of optimum allocation of resources," *Operations Research,* Vol. 11, No. 3, pp. 399-417 (1963); Section 11.2 is based on E. L. Lawler and M. D. Bell, "A method for solving discrete optimization problems," *Operations Research,* Vol. 14, pp. 1098-1112 (1966); Part of Section 11.4 is based on P. M. Ghare and R. E. Taylor, "Optimal redundancy for reliability," *Operations*

*Research,* Vol. 17, pp. 838-847 (1969). These are reprinted with permission of the Operations Research Society of America.

Section 11.3 is based on F. A. Tillman and J. M. Liittschwager, "Integer programming formulation of constrained reliability problems," *Management Science,* Vol. 13, No. 11, pp. 887-899 (1967). This is reprinted with the permission of the Institute of Management Sciences.

Frank A. Tillman
Ching-Lai Hwang
Way Kuo

# OPTIMIZATION OF SYSTEMS RELIABILITY

Reliability engineering appeared on the scene in the late 1940's and early 1950's and was first applied to the fields of communication and transportation.  Much of the early reliability work was confined to making tradeoffs between certain performance and reliability aspects of systems.  However, in today's highly complex systems, reliability has become increasingly important.

One goal of the reliability engineer is to find the best way to increase systems reliability.  Six important methods for doing this are [7]:  (1)  Keep the system as simple as is compatible with the performance requirements.  Nonessential components and unnecessarily complex configurations only increase the probability of system failure.  (2)  Increase the reliability of the components in the system.  (3)  Use parallel redundancy for the less reliable components or stages.  (4)  Use standby redundancy which is switched to active components when failure occurs.  (5)  Use repair maintenance where failed components are replaced but not automatically switched in as in (4).  (6)  Use preventive maintenance such that components are replaced by new ones whenever they fail or at some fixed interval, whichever comes first [2].

There are a number of measures to indicate the performance of equipment.  The most popular are reliability, availability and system effectiveness.  Reliability is simply the probability of successful operation, whereas availability is the probability that the system is operational and available, when it is needed.  To combine these measures into a single index, system effectiveness is often used. It is the overall capability of a system to accomplish its mission and is the product of its probability of reliability and availability. In this book we will concentrate on ways to improve system reliability through the optimal allocation of redundancies to systems which may be subject to constraints.

Reliability as stated before is defined to be the probability of successful operation.  One definition reads "Reliability is the probability that a system will perform satisfactorily for at least a given period of time when used under stated conditions" [1].  Therefore, the probability that a system successfully performs as designed is called "system reliability," sometimes called the "probability of survival."

System reliability is a measure of how well a system meets its design objective, and it is usually expressed in terms of the reliabilities of the subsystems or components.  The following terminologies are defined.  A "part" or "element" is the least subdivision of a system, or an item that cannot ordinarily be disassembled without being destroyed.  A "circuit" is a collection of parts that has a specific function.  A "component" then is a collection of parts and/or circuits which represents a self-contained element of a complete operating system and performs a function necessary to the operation of that system.  "Unit," "component," and "subsystem" are synonymous.  A "system" can then be characterized as a group of subsystems especially integrated to perform a specific operational function or functions.

Recall that the "reliability" of a system is the probability of the successful operation of the system for a specified period of time.  In describing the reliability of a given system it is necessary to specify (1) the equipment failure process, (2) the system configuration which describes how the equipment is connected and the rules of operation, and (3) the state in which the system is defined to be failed.  The equipment failure process describes the probability law governing those failures.  The system configuration, on the other hand, defines the manner in which the system reliability function will behave.  The third consideration in developing the reliability function for a nonmaintained system is to define the conditions of system failures.

A system in many cases is not confined to a single component. We want to suggest ways to evaluate the reliability of those systems which are simple as well as those which are extremely complex.  To develop functions expressing systems reliability, both conventional statistical theory and Markovian processes are used.

In this book a thorough discussion of reliability optimization techniques is presented.  A state-of-the-art review of the literature related to optimal system reliability with redundancy is presented in Chapter 2.  The literature is classified as follows.

Optimal system reliability models with redundancy:

    Series

    Parallel

    Series-parallel

    Parallel-series

    Standby

    Complex (nonseries, nonparallel)

Optimization techniques for obtaining optimal system configurations:

    Integer programming

    Dynamic programming

    Maximum principle

    Linear programming

    Geometric programming

    Sequential unconstrained minimization technique (SUMT)

    Modified sequential simplex pattern search

    Lagrange multipliers and Kuhn-Tucker conditions

    Generalized Lagrangian function

    Generalized reduced gradient (GRG)

    Heuristic approaches

      Parametric approaches

      Pseudo-Boolean programming

      Miscellaneous

A good deal of work has been done in the field of optimal redundancy allocation. However, to increase systems reliability, the system will spend more "cost" in weight, volume size, money expenditure, etc. to meet the requirement of higher systems reliability. Hence, in Chapters 3 through 12 optimization techniques are introduced to (a) maximize the systems reliability of various system configurations subject to the "cost" constraints, or (b) minimize any specific "cost" while satisfying the minimum requirement of system reliability.

The literature that has been published on optimal systems reliability is classified and reviewed. Various problems are also classified and solved by heuristic approaches, dynamic programming, and integer programming. These optimization techniques always give solutions of integer numbers which meet the implied integer requirement of redundancy allocation problems.

Sequential Unconstrained Minimization Technique (SUMT) has been widely used in solving many optimization problems. The Generalized Reduced Gradient method (GRG) and generalized Lagrangian functions method have been developed but have never been used in systems reliability optimization problems. The GRG method is an elaborate extension of hill-climbing gradient techniques, and has been coded in FORTRAN in a program named GREG. A new type of generalized or augmented Lagrangian function proposed by Sayama et al. for finding the solution of a nonlinear programming problem with inequality constraints is also explored in these chapters. Since none of these methods yields integer solutions, rounding-off procedures are applied whenever the redundancy allocation problem is solved.

The maximum principle, the method of Lagrange multipliers and the Kuhn-Tucker conditions, geometric programming, and several other optimization techniques (e.g. linear programming and separable programming) are also introduced.

In Chapter 13 a problem is presented which includes the determination of the optimal level of component reliability as well as the number of redundancies in each of the stages. The problem is one in which the component failure rates are variables and the optimal tradeoff between adding components in redundancy or the improvement of an individual component's reliability is considered. This becomes a mixed integer programming problem in which the system reliability is to be maximized as a function of component reliability level and the number of components used at each stage. The Hooke and Jeeves pattern search technique in combination with the heuristic approach by Aggarwal et al. is utilized to solve this problem.

Optimization problems also occur in connection with maintenance models. They have not been covered in this book. Note that surveys of literature on availability of maintenance models are presented in references 8 and 9.

## REFERENCES


1. Aeronautical Radio, Inc., *Reliability Engineering*, Englewood Cliffs, N. J.: Prentice-Hall (1964).

2. Goldman, A. S., and T. B. Slattery, *Maintainability: A Major Element of System Effectiveness*, New York: Wiley (1964).

3. Lloyd, D. K., and M. Lipow, *Reliability: Management, Methods, and Mathematics*, Englewood Cliffs, N. J.: Prentice-Hall (1962).

4. Polovko, A. M., *Fundamentals of Reliability Theory*, New York: Academic Press (1968).

5. Sandler, G. H., *System Reliability Engineering*, Englewood Cliffs, N. J.; Prentice-Hall (1963).

6. Shooman, M. L., *Probabilistic Reliability: An Engineering Approach*, New York: McGraw-Hill (1968).

7. Smith, C. O., *Introduction to Reliability in Design*, New York: McGraw-Hill (1976).

8.  Lie, C. H., C. L. Hwang, and F. A. Tillman, "Availability of maintained systems:  A state-of-the-art survey," *AIIE Transactions,* Vol. 9, pp. 247-259 (1977).

9.  Osaki, S., and T. Nakagawa, "Bibliography for reliability and availability of stochastic systems," *IEEE Trans. on Reliability,* Vol. R-25, No. 4, pp. 284-287 (1976).

     OPTIMIZATION TECHNIQUES FOR SYSTEMS RELIABILITY
WITH REDUNDANCY:  A REVIEW

## 2.1  INTRODUCTION

The reliable performance of a system for a mission under various conditions is of utmost importance in many industrial, military, and
everyday life situations.  Although the qualitative concepts of
reliability are not new, its quantitative aspects have been developed
over the past two decades.  Such development has resulted from
the increasing need for highly reliable systems and components with
more safety and less cost.

There exist several methods to improve systems reliability. Some
of these methods approach the problem by using large safety factors,
reducing the complexity of the system, increasing the reliability
of constituent components through a product improvement program,
using structural redundancy, or practicing a planned maintenance
and repair schedule.  A good deal of effort has been centered
in the field of optimal redundancy allocation.

A state-of-the-art review of the literature related to optimal
systems reliability with redundancy is presented in this chapter.
The first part of the reference list is concerned with basic reliability [1-17] and optimization techniques [18-66].  The references
for the various optimization techniques are:  optimization techniques
in general [18-23], integer programming [24-29], the maximum principle [30-33], the generalized reduced gradient method (GRG) [34-42],
modified sequential simplex pattern search [43-46], the sequential
unconstrained minimization technique (SUMT) [47-52], the method of

Lagrange multipliers and the Kuhn-Tucker conditions [53-54], the generalized Lagrangian functions method [55-58], dynamic programming [59-63], and geometric programming [64-66].

The second part of the reference list concentrates mainly  on articles relevant to the optimization of systems reliability with redundancy [67-143], which are classified into two categories: the system configurations and the optimization techniques employed. (See Tables 1 and 2.)  In Table 1, the literature for the different system configurations is separated into the following model sub-categories: series, parallel, series-parallel, parallel-series, standby, and nonseries-nonparallel models.  In Table 2, the same literature is reclassified to indicate the variety of optimization techniques utilized.

Although we have tried to give a reasonably complete survey, those papers not included were either inadvertently overlooked or considered not to bear directly on the topics of this survey. We apologize to both the readers and the researchers if we have omitted any relevant papers.


## 2.2    SYSTEMS MODELS

In this review, we assume that the reader is familiar with the material treated in these models.  For a discussion of the definitions and formulations of the basic concepts, we suggest reviewing the books on reliability as stated in the references [1-17].  We will briefly describe each of the models considered in this survey.

The first model considered is an N-stage series system and is shown in Fig. 1.  In this system, the functional operation depends upon the proper operation of all system components.  If one component fails, the system fails.  The second model is an M-stage parallel system which is shown in Fig. 2.  There are M paths connecting the input to the output, and all components must fail for the system to fail.

Table 1    The Reference Classification for the Optimization of
Systems Reliability with Redundancy with Regard to Various System
Configuration.

| SYSTEM CONFIGURATION | REFERENCES |
|---|---|
| Series | 70, 72, 76, 77, 78, 79, 83, 87, 88, 89, 108, 122, 125, 129, 130, 136, 139, 141, 142 |
| Parallel | 70, 71, 79, 88, 89, 95, 104, 122, 129, 136, 139, 141, 142 |
| Series-Parallel | 79, 88, 89, 94, 95, 110, 122, 124, 129, 130, 141, 142 |
| Parallel-series | 62, 69, 72, 75, 77, 78, 79, 80, 81, 82, 83, 84, 87, 88, 89, 92, 93, 94, 96, 98, 100, 101, 104, 105, 106, 107, 108, 109, 110, 111, 112, 114, 115, 116, 117, 118, 119, 120, 122, 124, 126, 127, 128, 129, 130, 132, 134, 135, 136, 137, 138, 139, 141, 142, 143 |
| Standby | 77, 89, 90, 93, 108, 120, 121, 122, 127, 129, 130, 135, 141, 142 |
| Nonseries-nonparallel (including bridge network) | 68, 70, 71, 73, 79, 80, 91, 97, 136, 140 |

Table 2    The Reference Classification for the Optimization Techniques
Employed for Systems Reliability with Redundancy.

| OPTIMIZATION TECHNIQUE | REFERENCES |
| --- | --- |
| Integer programming | 86, 87, 90, 92, 98, 103, 105, 106, 107, 112, 116, 120, 135, 137, 139 |
| Dynamic programming | 75, 80, 84, 94, 96, 99, 101, 104, 108, 127, 128, 133, 143 |
| The maximum principle | 82, 110, 138 |
| Linear programming | 98, 131 |
| Geometric programming | 83, 114 |
| Sequential unconstrained minimization technique | 91, 134, 136, 140 |
| Modified sequential simplex pattern search | 71, 118 |
| The method of Lagrange multipliers and the Kuhn-Tucker conditions | 77, 78, 81, 108, 110, 111, 133 |
| The generalized Lagrangian function method | 100 |
| Generalized reduced gradient method | 100 |
| Heuristic approach | 68, 69, 100, 132 |
| Parametric approach | 71, 72, 73, 87 |
| Pseudo-Boolean programming | 93 |
| Others (miscellaneous) | 76, 77, 78, 88, 89, 95, 100, 109, 115, 117, 119, 121, 122, 125, 129, 130, 141, 142 |

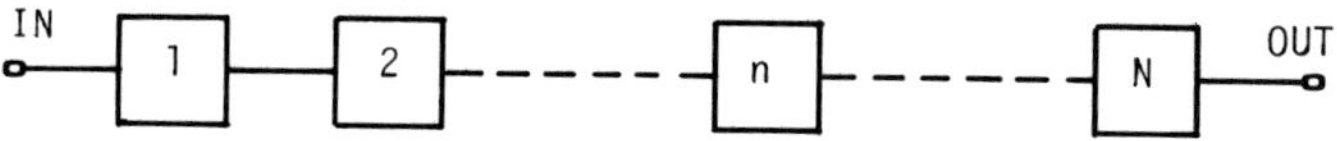

Fig. 1:  An N-stage series system

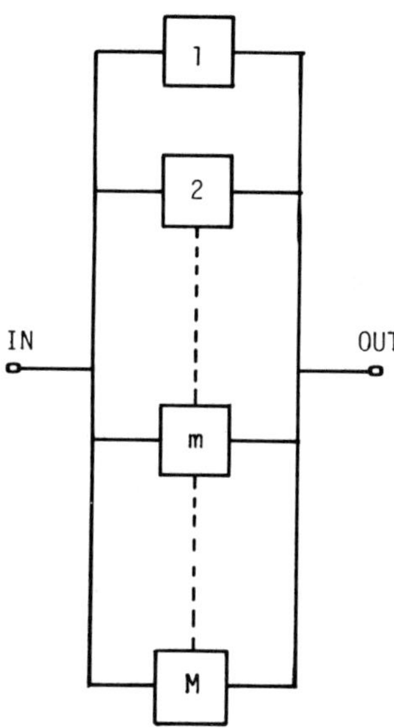

Fig. 2:  An M-Stage parallel system

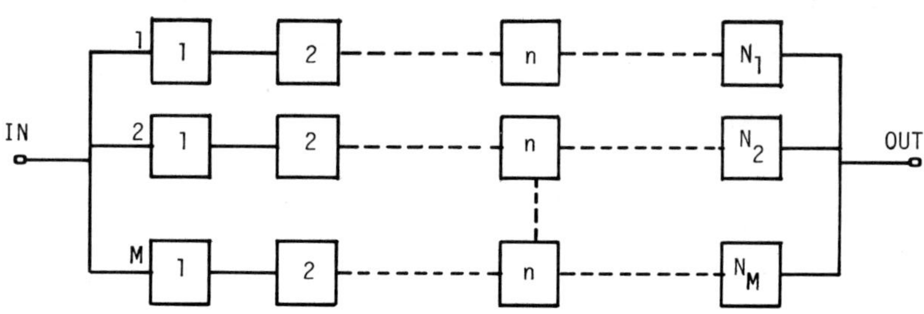

Fig. 3:  A mixed series-parallel system

Figure 3 shows a mixed series-parallel system in which N components are connected in a series arrangement where M such series connections are connected in parallel to form the system. Figure 4 shows a mixed parallel-series system. In this system, N stages are connected in series where components are connected in parallel at each stage.

An element standby system is shown in Fig. 5, which has the same form as a mixed parallel-series system. However, in this system the parallel components are not all active at the same time. Figure 6 shows a system standby system which has the same form as a mixed series-parallel system. However, when a system standby system is used, the parallel M series subsystems are not all active at the same time.

Figure 7 shows a typical nonseries-nonparallel reliability system. The reliability of this system can be evaluated by using either conditional probabilities or other approaches. Figure 8 shows a complex bridge network system, one of the complex reliability systems in the form of a bridge network.

Table 1 presents the literature on the optimization of systems reliability for the above systems models.

## 2.3 STATEMENT OF THE VARIOUS OPTIMIZATION PROBLEMS

The structure of the optimization problems, which are relevant to our survey, are stated below and the literature is identified in Table 3.

For an N-stage series model (see Fig. 1), the problem is one of allocating the reliability to each of the components so that the reliability is maximized. This can be stated as

*Problem 1.*    Maximize:

$$R_s = \prod_{j=1}^{N} R_j$$

subject to

$$\sum_{j=1}^{N} g_{ij}(R_j) \le b_i, \qquad i = 1, 2, \ldots, m$$

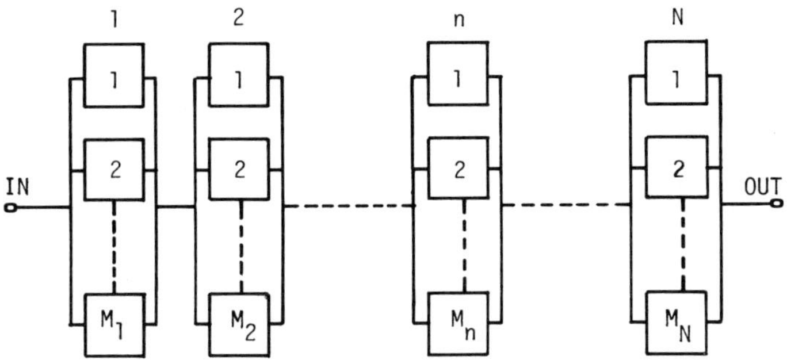

Fig. 4:   A mixed parallel-series system

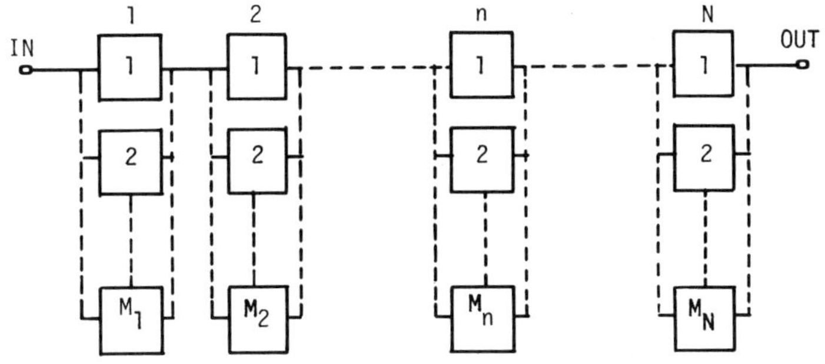

Fig. 5:   An element standby system

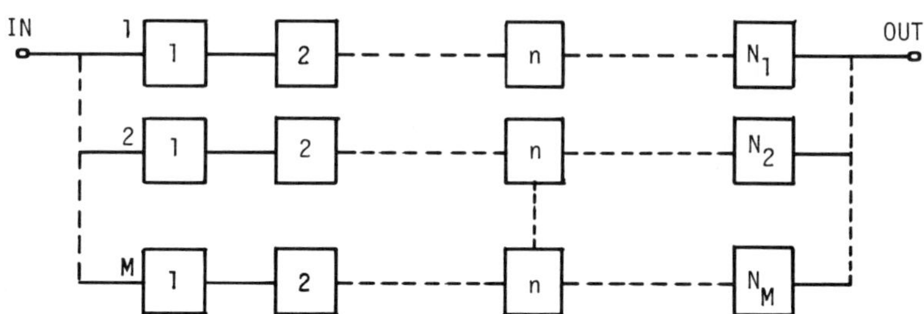

Fig. 6:   A standby system

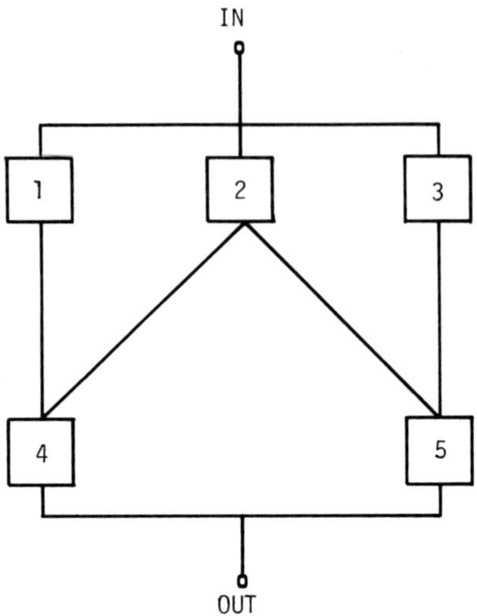


Fig. 7:   A typical nonseries-nonparallel reliability system

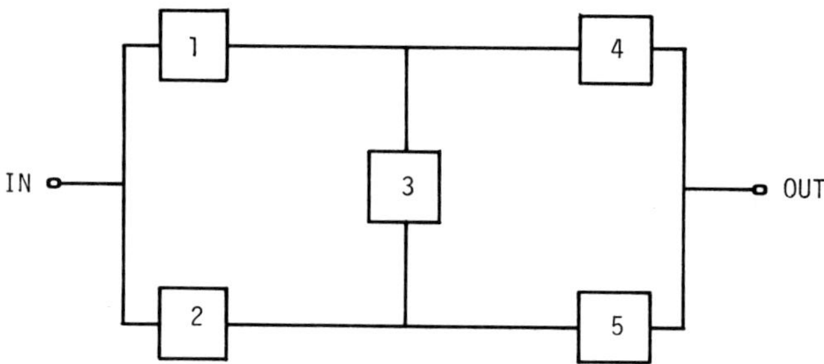


Fig. 8:   A complex bridge network system

Table 3    The Reference Classifications with Regard to the Structure
of the Optimization Problems.

| FORMULATION OF PROBLEMS | REFERENCES |
|---|---|
| Problem 1:  Optimum reliability allocation **for an N-stage** series system | 9, 76, 99, 100, 110, 113, 125 |
| Problem 2:  Optimum redundancy allocation, maximization of systems reliability subject to cost constraints | 69, 72, 75, 77, 80, 81, 83, 84, 88, 89, 90, 92, 93, 94, 96, 98, 100, 105, 106, 107, 108, 109, 110, 111, 112, 113, 114, 115, 116, 117, 118, 119, 120, 121, 122, 123, 126, 129, 130, 132, 134, 135, 136, 137, 138, 139, 141, 142, 143, |
| Problem 3:  "Cost" minimization problems subject to the minimum requirement of systems reliability | 72, 78, 90, 91, 93, 100, 112, 113, 122, 128, 131, 135, 137, 139 |
| Problem 4:  Systems reliability maximization for a nonseries-nonparallel system | |
|     Reliability allocation | 80, 91, 100, 140 |
|     Redundancy allocation | 68, 71, 136 |
| Others | |
|     Maximization of the system profit | 82, 128 |
|     Maximization of the ratio of systems reliability to the power demand of the system | 104, |

where $R_s$ is the systems reliability, $R_j$ is the component reliability of the $j^{th}$ stage, $g_{ij}(R_j)$ is the resource i consumed at stage j, and $b_i$ is the total amount of resource i available.  The function, $g_{ij}(R_j)$, can be either linear or nonlinear with respect to the component reliability, $R_j$.

A further delineation of this problem can be stated as finding the optimum number of redundancies (see Figs. 2-6) which maximize the system reliability subject to "cost" constraints, or the minimization of system costs subject to the condition that the system reliability is equal to or greater than a desired level. These problems are stated as

*Problem 2.*   Maximize

$$R_s = \prod_{j=1}^{N} R_j(X_j)$$

subject to

$$\sum_{j=1}^{N} g_{ij}(X_j) \leq b_i, \qquad i = 1,2,\ldots,m$$

where $R_j$ is the reliability of the $j^{th}$ stage (subsystem), and is a function of the number of components in each stage, $X_j$.

*Problem 3.*   Minimize

$$C_s = \sum_{j=1}^{N} C_j(X_j)$$

subject to

$$R_s = \prod_{j=1}^{N} R_j(X_j) \leq R_r$$

where $C_s$ is the total cost of the system and $C_j$ is the cost of the jth stage which is a function of the number of components in each stage, $X_j$ .  The systems reliability, $R_s$ , has to be greater than or equal to the required systems reliability, $R_r$ .

The system reliability for complex systems (see Figs. 7 and 8) is obtained by using either conditional probabilities or other network approaches.  The optimization problem is stated as Problem 4.

*Problem 4.*  Maximize

$$R_s = f(R_1, R_2, \ldots, R_N)$$

subject to

$$\sum_{j=1}^{N} g_{ij}(R_j) \le b_i, \qquad i = 1, 2, \ldots, m$$

where the systems reliability is a function of the component reliability, $R_j$.

## 2.4  OPTIMIZATION TECHNIQUES USED TO DETERMINE THE OPTIMAL SYSTEMS RELIABILITY

Most of the problems stated are nonlinear integer programming problems. They are more difficult to solve than general nonlinear programming problems because their solutions must be integers.  Many algorithms have been proposed, but only a few have proven effective when applied to large-scale nonlinear programming problems.  None have proven to be superior over the others so that it could be classified as the algorithm for solving general nonlinear programming problems [22].

The literature on the optimization techniques which are relevant to this survey is classified and presented in Table 2. Each of the optimization techniques employed in the 77 papers [67-143] has limited success in solving all of the problems.

Although integer programming [24-29] yields integer solutions, the transformation of nonlinear objective functions and constraints into linear forms so that integer programming can be applied is a difficult task.  In addition, the various integer programming techniques do not guarantee that optimal solutions can be obtained in a reasonable time.  Dynamic programming [59-63] has the dimensionality difficulties which increase with the increase of the number of state variables, and it is hard to solve problems with more than three constraints.  Similarly, the maximum principle [30-33] has difficulty in solving problems with more than three constraints.

Likewise geometric programming [64-66] is restricted to problems that
can be formulated by posynomial functions.

The sequential unconstrained minimization technique (SUMT) [47-
52], the generalized reduced gradient method (GRG) [34-42], the modified sequential simplex pattern search [43-46], and the generalized
Lagrangian function method [55-58] are probably the few techniques
demonstrated effective when applied to large-scale nonlinear programming problems.  However, the solutions are nonintegers and hence the
optimal solution, which must be an integer, is not guaranteed.

## 2.5   REMARKS ON NEW PROBLEMS

All the optimization techniques employed in the papers surveyed have
limited success in solving some small-scale systems reliability optimization problems.  Few techniques have been demonstrated effective
when applied to large-scale problems.

There are some new directions in which additional optimization
work would be fruitful.  For example, one extension to the usual
reliability optimization problem is to include the determination
of the optimal level of component reliability and the number of redundancies  in each of the stages simultaneously.  The problem is one
in which the component failure rate is a variable and the optimal
trade-off between  adding components in redundancy or improvement
of the individual component reliability is to be determined (see
Chapter 13). Another example is one where the optimization of multistage systems reliability is achieved by choosing a more reliable
component out of several possible candidates at stage 1, adding redundant components in parallel at stage 2, and using a k-out-of-n:
G configuration at stage 3 (see Example 3-4).

Cost data for improving systems reliability is critically needed;
very little is presently available.  To formulate objective function
and constraints, actual cost data is necessary to realistically model
the problems.

Increasing complexity of modern-day equipment, both in military
and commercial areas, has brought with it new engineering problems
involving high performance, reliability, and maintainability.  In
this regard availability, which is a combined measure of maintain-

ability and reliability, has received wide increased usage as a
measure of systems effectiveness.  A survey of this literature is
presented in reference [144] along with a logical classification
of various aspects of this problem.

## 2.6   OPTIMIZATION OF SYSTEMS RELIABILITY

Optimization techniques have their inherent characteristics and specific superiorities to solve general linear or nonlinear programming
problems.  In the following chapters, various optimization techniques
are presented to
(1) maximize systems reliability by adding the redundant
    components in each specified subsystem,
(2) maximize systems reliability by choosing a suitable stage
    reliability in each specified subsystem,
(3) minimize the "cost" of the system while satisfying the
    minimum requirement of systems reliability, or
(4) minimize the "cost" of a multi-function system while
    satisfying the minimum requirement of each individual
    system reliability.
"Cost" constraints of cost, weight, volume, or some combination of
these factors are imposed on a system with series, parallel, or complex configuration.  Each of the constraint functions is an increasing
function of the component reliability and/or number of components
used at each stage.  Various "cost" functions are used.

In this chapter, references for optimization techniques for
systems reliability with redundancy have been reviewed.  The computational procedures of the optimization techniques, which have or
have not been applied in the optimization of systems reliability,
will be described in the following chapters.  These optimization
techniques are:
1.  heuristic approach
2.  dynamic programming
3.  the discrete maximum principle
4.  the sequential unconstrained minimization technique (SUMT)
5.  the generalized reduced gradient method (GRG)

    6.  method of Lagrange multipliers and the Kuhn-Tucker
        conditions
    7.  the generalized Lagrangian functions method
    8.  geometric programming
    9.  integer programming
   10.  others (a classical approach, parametric method, linear
        programming, and separable programming)

Among these optimization techniques, both GRG and the generalized
Lagrangian functions method are very promising. Heuristic approaches
and dynamic programming, having been successfully applied to the
redundancy allocation problems, will be critically reviewed, classified, and modified. To present a comprehensive discussion, the other
optimization techniques will also be used to solve various reliability
optimization problems. Before dealing with each specific technique,
the following assumptions are made:

(1)  Each subsystem is considered essential for the overall operational
     success of the mission, if all the subsystems are operational
     in series.

(2)  All the subsystems in series, parallel, or complex configuration
     are $\underline{s}$-independent*; also there are statistically independent
     parallel redundant components in each subsystem. In parallel
     redundancy, all units have the same risk of failure (or success)
     whether they are spares or active.

(3)  Before the requirement of linearization for some specific
     optimization techniques, the constraints of "cost" do not need
     to be in a linear form.

(4)  Good/bad is a sufficient description for each component, sub-
     system, and the whole system. In parallel cases, unless spec-
     ified, only one component needs to be good for the subsystem
     to be good, this is considered to be a 1-out-of m: G configuration.
     No assumption is made about the hazard rates of the components,
     except that they are reflected in the reliability of the com-
     ponents.

---

*$\underline{s}$-independent:  $\underline{s}$ implies "statistical(ly).".

(5) Without the specific optimization knowledge of the mission requirements, realistic decisions on redundancy, design change, and other aspects of reliability improvement cannot be reached. Tradeoffs can be considered only between optimal redundancy components and "cost" measures.

(6) The constraints are additive between subsystems.

(7) The redundant models are based on the assumption that individual component or path failure has no effect on the operation of the surviving paths.

(8) The connection nodes may accrue some "cost", but are assumed to function perfectly given the system is working.

## REFERENCES    (1) ON RELIABILITY AND OPTIMIZATION TECHNIQUES

### RELIABILITY


1. ARINC Res. Corp., *Reliability Engineering*, Englewood Cliffs, N. J.: Prentice-Hall (1964).

2. Amstadter, B. L., *Reliability Mathematics*, New York:  McGraw-Hill (1971).

3. Barlow, R. E., and F. Proschan, *Mathematical Theory of Reliability*, New York:  Wiley (1965).

4. Barlow, R. E., and F. Proschan, *Statistical Theory of Reliability and Life Testing*, New York:  Holt, Rinehart, and Winston (1975).

5. Bazovsky, I., *Reliability Theory and Practice*, Englewood Cliffs, N. J.:  Prentice-Hall (1961).

6. Calabro, S. R., *Reliability Principles and Practice*, New York: McGraw-Hill (1962).

7. Gnedenko, B. V., Y. K. Belyayer, and A. D. Solovyev, *Mathematical Methods of Reliability Theory*, New York: Academic Press (1969).

8. Grouchko, D. (ed.), *Operations Research and Reliability*, New York:  Gordon and Breach (1969).

9.  Lloyd, D. K., and M. Lipow, *Reliability: Management, Methods, and Mathematics,* Englewood Cliffs, N. J.: Prentice-Hall (1962).

10. Mann, N. R., R. E. Shafer, and N. D. Singpurwalla, *Methods for Statistical Analysis of Reliability and Life Data,* New York: Wiley (1974).

11. Meyer, P. L., *Introductory Probability and Statistical Application,* Reading, Mass.: Addison-Wesley (1966).

12. Myers, R. H., K. L. Wong, and H. M. Gordy, *Reliability Engineering for Electronic System,* New York: Wiley (1964).

13. ·Pieruschaka, E., *Principles of Reliability,* Englewood Cliffs, N. J.: Prentice-Hall (1963).

13a. Polovko, A. M., *Fundamentals of Reliability Theory,* New York: Academic Press (1968).

14. Rau, J. G., *Optimization and Probability in Systems Engineering,* New York: Van Nostrand Reinhold (1970).

15. Sandler, G. H., *System Reliability Engineering,* Englewood Cliffs, N. J.: Prentice-Hall (1963).

16. Shooman, M. L., *Probabilistic Reliability: An Engineering Approach,* New York: McGraw-Hill (1968).

17. Zelen, M. (ed.), *Statistical Theory of Reliability,* Madison, University of Wisconsin Press (1964).



## GENERAL OPTIMIZATION TECHNIQUES



18. Abadie, J. (ed.), *Integer and Nonlinear Programming,* New York: North-Holland/American Elsevier (1970).

19. Beveridge, G. S. G., and R. S. Schechter, *Optimization: Theory and Practice,* New York: McGraw-Hill (1970).

20. Fletcher, R., *Optimization,* New York: Academic Press (1969).

21. Leitmann, G., *Optimization Techniques with Applications to Aerospace Systems,* New York: Academic Press (1962).

22. Himmelblau, D. M., *Applied Nonlinear Programming,* New York: McGraw-Hill (1972).

23. Wilde, D. J., *Optimum Seeking Methods,* Englewood Cliffs, N. J.: Prentice-Hall (1964).

## INTEGER PROGRAMMING

24. Garfinkel, R. S., and G. L. Nemhauser, *Integer Programming,* New York: Wiley (1972).

25. Geoffrion, A. M., "An inproved implicit enumeration approach for integer programming," RAND Corp. Rept. RM-5644-PR June 1968, or Oper. Res. Vol. 17, pp. 437-454 May-June (1969).

26. Gomony, R. E., "Outline of an algorithm for integer solutions to linear problem," Bulletin of the American Mathematical Society, Vol. 64, No. 5, pp. 275-278 (1958).

27. Hu, T. C., *Integer Programming and Network Flows,* Reading, Mass: Addison-Wesley (1969).

28. Salkin, H. M., *Integer Programming,* Reading, Mass: Addison-Wesley (1975).

29. Taha, H. A., *Integer Programming: Theory, Applications and Computations,* New York: Academic Press (1975).


## THE MAXIMUM PRINCIPLE

30. Denn, M. M., *Optimization by Variational Methods,* New York: McGraw-Hill (1969).

31. Fan, L. T., and C. S. Wang, *The Discrete Maximum Principle: A Study of Multistage Systems Optimization,* New York: Wiley (1964).

32. Fan, L. T., *The Continuous Maximum Principle: A Study of Complex Systems Optimization,* New York: Wiley (1966).

33. Pontryagin, L. S., V. G. Boltyanskii, R. V. Gamkrelidge, and E. F. Mischenko, *The Mathematical Theory of Optimal Processes,* trans. K. N. Trirogoff, New York: Wiley-Interscience (1962).


## GENERALIZED REDUCED GRADIENT METHOD

34. Abadie, J. and J. Guigou, Numerical experiments with the GRG method, in *Integer and Nonlinear Programming,* ed. J. Abadie, Amsterdam: North-Holland (1970).

35. Abadie, J., *Solution des questions de degenerscence dans la methode GRG,* EDF* note HI 143/100 due 25 September (1969).

36. Abadie, J., Application of the GRG Algorithm to Optimal Control Problem, in *Integer and Nonlinear Programming,* ed. J. Abadie, Amsterdam: North-Holland (1970).

37.  Abadie, J., and J. Carpenter, Generalization of the Wolfe Reduced Gradient Method to the Case of Nonlinear Constraints, in *Optimization*, ed. R. Fletcher, New York:  Academic Press (1969).

38.  Abadie, J. and J. Guigou, Gradient reduit generalise, EDF* note HI 069/2 du Avril (1969).

39.  Colvill, A. R., A comparative study of nonlinear programming codes, IBM New York Scientific Center, Report 320-2949 (1968).

40.  Guigou, J., Presentation et utilisation du code GRG, EDF* note Hl 102/02 du 9 Juin (1969).

41.  Guigou, J., Presentation et utilisation de code GREG, EDF* note Hl 582/2 du 25 Mai (1971).

42.  Hwang, C. L., J. L. Williams, and L. T. Fan, *Introduction to the Generalized Reduced Gradient Method*, Institute for Systems Design and Optimization, Report No. 39, Kansas State University, Manhattan (1972).

## MODIFIED SEQUENTIAL SIMPLEX PATTERN SEARCH

43.  Box, M. J., "A new method of constrained optimization and a comparison with other methods," *Computer Journal*, Vol. 8, pp. 42-52 (1965).

44.  Fan, L. T., C. L. Hwang, and F. A. Tillman, "A sequential simplex pattern search solution to production planning problems," *AIIE Transactions*, Vol. 1, pp. 267-273 (1969).

45.  Nelder, J. A., and R. Mead, "A simplex method for function minimization," *Computer Journal*, Vol. 7, pp. 308-313 (1965).

46.  Sheela, B. V., and P. Ramamoorthy, "SWIFT-A new constrained optimization technique," *Computer Methods in Applied Mechanics and Engineering*, Vol. 6, pp. 309-317 (1975).

## SEQUENTIAL UNCONSTRAINED MINIMIZATION TECHNIQUE

47.  Bracken, J., and G. P. McCormick, *Selected Applications of Nonlinear Programming*, New York:  Wiley (1968).

48.  Fiacco, A. V., and G. P. McCormick, *Nonlinear Programming: Sequential Unconstrained Minimization Techniques*, New York: Wiley (1968).

49. Hwang, C. L., K. C. Lai, F. A. Tillman, and L. T. Fan, "Optimization of system reliability by the sequential unconstrained minimization techniques," *IEEE Transactions on Reliability,* Vol. R-24, pp. 133-135 (1975)

50. Kowalik, J., and M. R. Osborne, *Methods for Unconstrained Optimization Problems,* New York: Elsevier (1968).

51. Lai, K. C., *Optimization of Industrial Management Systems by the Sequential Unconstrained Minimization Technique,* A master's report, Kansas State University, Manhattan (1970).

52. McCormick, G. P., W. C. Mylander, III, and A. V. Fiacco, "RAC Computer Program Implementing the Sequential Unconstrained Minimization Technique for Nonlinear Programming," SHARE Number 3189.

## METHOD OF LAGRANGE MULTIPLIERS AND THE KUHN-TUCKER CONDITIONS

53. Hwang, C. L., P. K. Gupta, and L. T. Fan, *Method of Lagrange Multipliers and the Kuhn-Tucker Conditions,* Institute for Systems Design and Optimization, Report No. 60, Kansas State University, Manhattan (1974).

54. Kuhn, N. W. and A. W. Tucker, Non-Linear Programming, *Proceedings of the Second Berkeley Symposium on Mathematical Statics and Probability,* J. Neyman, Ed., Berkeley: University of California Press, pp. 481-492 (1951).

## THE GENERALIZED LAGRANGIAN FUNCTIONS METHOD

55. Everett, H., "Generalized Lagrange Multiplier Method for Solving Problems of Optimal Allocation of Resources," *Operations Research,* vol. 11, pp. 399-417 (1963).

56. Lootsman, F. A., "A Survey of Methods for Solving Constrained Minimization Problems iva Unconstrained minimization," in *Numerical Methods for Non-Linear Optimization,* F. A. Lootsma, (ed.) Academic Press (1973); also in *Optimization and Design,* M. Avriel, M. J. Rijckaert, and P. J. Wilde (eds.) Englewood Cliffs, N. J.: Prentice-Hall, pp. 88-118 (1973).

57. Roode, J. D., "Generalized Lagrange Function and Mathematical Programming," in *Optimization,* ed. R. Fletcher, New York: Academic Press (1969).

58. Sayama, H., Y. Kameyama, H. Nahayama, and Y. Sawaragi, *The Generalized Lagrangian Function for Mathematical Programming Problems,* Institute for Systems Design and Optimization, No. 55, Kansas State University, Manhattan (1974).

DYNAMIC PROGRAMMING

59. Bellman, R., *Dynamic Programming,* Princeton, N. J.:  Princeton University Press (1957).

60. Bellman, R., and S. E. Dreyfus, *Applied Dynamic Programming,* Princeton, N. J.: Princeton University Press (1962).

61. Bellman, R., *Modern Analytic and Computational Methods in Science and Mathematics,* New York:  Elsevier (1968).

62. Hadley, G., *Nonlinear and Dynamic Programming,* Reading, Mass: Addison-Wesley (1964).

63. Nemhauser, G. L., *Introduction to Dynamic Programming,* New York: Wiley (1967).


GEOMETRIC PROGRAMMING

64. Duffin, R. J., E. L. Peterson, and C. Zener, *Geometric Programming,* New York:  Wiley (1967).

65. Nijkamp, P., *Planning of Industrial Complexes by Means of Geometric Programming,* Rotterdam, Netherlands:  Rotterdam University Press (1972).

66. Zener, C., *Engineering Design by Geometric Programming,* New York: Wiley-Interscience, (1971).


(2)  ON OPTIMIZATION OF SYSTEMS RELIABILITY WITH REDUNDANCY

67. Aggarwal, K. K., and J. S. Gupta, "On minimizing the cost of reliable systems," *IEEE Transactions on Reliability,* Vol. R-24, No. 3, pp. 205 (1975).

68. Aggarwal, K. K., "Redundancy optimization in general systems," *IEEE Transactions on Reliability,* Vol. R-25, No. 5, pp. 330-332 (1976).

69. Aggarwal, K. K., J. S. Gupta, and K. B. Miscro, " A new heuristic criterion for solving a redundancy optimization problem," *IEEE Transactions on Reliability,* Vol. R-24, No. 1, pp. 86-87 (1975).

70. Aggarwal, K. K., and K. B. Misra, and J. S. Gupta, "Reliability evaluation-a comparative study of different techniques," *Microelectronics and Reliability,* Vol. 14, No. 1, pp. 49-56, (1975).

72. Banerjee, S. K., and K. Rajamani, "Optimization of system reliability using a parametric approach," *IEEE Transactions on Reliability,* Vol. R-22, pp. 35-39 (1973).

73. Banerjee, S. K., and K. Rajamani, "Parametric representation of probability in two dimensions - a new approach in system reliability evaluation," *IEEE Transactions on Reliability,* Vol. R-21, pp. 56-60 (1972).

74. Banerjee, S. K., and K. Rajamani, "Closed form solutions for delta-star and star-delta conversions of reliability networks," *IEEE Transactions on Reliability,* Vol. R-25, No. 2, pp. 118-119 (1976).

75. Bellman, R. E., and S. E. Dreyfus, "Dynamic programming and reliability of multicomponent devices," *Operations Research,* Vol. 6, pp. 200-206 (1958).

76. Beraha, D. and K. B. Misra, "Reliability optimization through random search algorithm," *Microelectronics and Reliability,* Vol. 13, pp. 295-297 (1974).

77. Black, G., and F. Proschan, "On optimal redundancy," *Operations Research,* Vol. 7, pp. 581-588 (1959).

78. Bodin, L. D., "Optimization procedure for the analysis of coherent structures," *IEEE Transactions on Reliability,* Vol. R-18, No. 3, pp. 118-126 (1969).

79. Brown, D. B., "A computerized algorithm for determining the reliability of redundant configurations," *IEEE Transactions on Reliability,* Vol. R-20, No. 3, pp. 121-124, (1971).

80. Burton, R. M., and G. T. Howard, "Optimal system reliability for a mixed series and parallel structure," *Journal of Mathematical Analysis and Applications,* Vol. 28, pp. 370-382 (1969).

81. Everett, H. III, "Generalized Lagrange multiplier method for solving problems of optimum allocation of resources," *Operations Research,* Vol. 11, No. 3, pp. 399-417 (1963).

82. Fan, L. T., C. S. Wang, F. A. Tillman, and C. L. Hwang, "Optimization of systems reliability," *IEEE Transactions on Reliability,* Vol. R-16, pp. 81-86 (1967).

83. Federowicz, A. J., and M. Mazumdar, "Use of geometric programming to maximize reliability achieved by redundancy," *Operations Research,* Vol. 16, No. 5, pp. 948-954 (1968).

84. Fyffe, D. E., W. W. Hines, and N. K. Lee, "System reliability allocation and a computational algorithm," *IEEE Transactions on Reliability,* Vol. R-17, No. 2, pp. 64-69 (1968).

85.   Geisler, M. A., and H. W. Karr, "The design of military supply
      tables for spare parts," *Operations Research,* Vol. 4, pp. 431-
      444 (1956).

86.   Gen, M., H. Okuno, and S. Shinofuji, "An optimizing method in
      system reliability with failure-modes by implicit enumeration
      algorithm," *J. of the Operations Research of Japan,* Vol. 19,
      pp. 99-116 (1976).

87.   Ghare, P. M., and R. E. Taylor, "Optimal redundancy for reli-
      ability in series system," *Operations Research,* Vol. 17, pp. 838-
      847 (1969).

88.   Gordon, R., "Optimum component redundancy for maximum system
      reliability," *Operations Research,* Vol. 5, pp. 229-243 (1957).

89.   Hees, Ir. R. N. v., and Ir. H. W. v. d. Meerendonk, "Optimal
      reliability of parallel multi-component systems," *Operations
      Research Quarterly,* Vol. 12, No. 1, pp. 16-26 (1961).

90.   Hwang, C. L., L. T. Fan, F. A. Tillman, and S. Kumar, "Optimi-
      zation of life support system reliability by an integer program-
      ming method,"*AIIE Transactions,* Vol. 3, No. 3, pp 229-238 (1971).

91.   Hwang, C. L., K. C. Lai, F. A. Tillman, and L. T. Fan, "Opti-
      mization of system reliability of the sequential unconstrained
      minimization techniques," *IEEE Transactions on Reliability,*
      Vol. R-24, pp. 133-135 (1975).

92.   Hyun, K. N., "Reliability optimization by 0-1 programming for
      a system with several failure modes," *IEEE Transactions on Reli-
      ability,* Vol. R-24, No. 3, pp. 206-210 (1975).

93.   Inoue, K., S. L. Gandi, and E. J. Henley, "Optimal reliability
      design of process systems," *IEEE Transactions on Reliability,*
      Vol. R-23, No. 1, pp. 29-33 (1974).

94.   Jensen, P. A., "Optimization of series-parallel-series networks,"
      *Operations Research,* Vol. 18, pp. 471-482 (1970).

95.   Kneale, S. G., "Reliability of parallel systems with repair and
      switching," *Proceedings of the Seventh National Symposium on
      Reliability and Quality Control,* pp. 129-133 (1961).

96.   Kettelle, J. D., "Least-cost allocation of reliability investment,"
      *Operations Research,* Vol. 10, pp. 249-265 (1967).

97.   Kim, Y. H., K. E. Case, and P. M. Glare, "A Method for computing
      complex system reliability," *IEEE Transactions on Reliability,*
      Vol. R-21, No. 4, pp. 215-219 (1972).

98. Kolesar, P. J., "Linear programming and the reliability of multi-component systems," *Naval Research Logistics Quarterly,* Vol. 15, pp. 317-327 (1967).

99. Kulshrestha, D. K., and M. C. Gupta, "Use of dynamic programming for reliability engineers," *IEEE Transactions on Reliability,* Vol. R-22, pp. 240-241 (1973).

100. Kuo, W., "Optimization Techniques for Systems Reliability with Redundancy," M. S. Thesis, Kansas State University, Manhattan, (1977).

101. Lambert, B. K., A. G. Walvekar, and J. P. Hirmas, "Optimal redundancy and availability allocation in multistage systems," *IEEE Transactions on Reliability,* Vol. R-20, No. 3, pp. 182-185 (1971).

102. Laut, S., "Subsystem optimization effectiveness improvement by the option tradeoff analysis process," *IEEE Transactions on Systems Science and Cybernetics,* Vol. SSC-4, No. 2, pp. 133-137 (1968).

103. Lawler, E. L., and M. D. Bell, " A method for solving discrete optimization problems," *Operations Research,* Vol. 14, pp. 1098-1112 (1966).

104. Liittschwager, J. M., "Dynamic programming in the solution of a multistage reliability problem," *Journal of Industrial Engineering,* Vol. 15-16, pp. 168-175 (1964-1965).

105. Luus, R., "Optimization of system reliablity by a new nonlinear integer programming procedure," *IEEE Transactions on Reliability,* Vol. R-24, pp. 14-16 (1975).

106. McLeavey, D. W., "Numerical investigation of parallel redundancy in series systems," *Operations Research,* Vol. 22, pp. 1110-1117 (1974).

107. McLeavey, D. W., and J. A. McLeavey, "Optimizations of system reliability by branch-and bound," *IEEE Transactions on Reliability,* Vol. R-25, No. 5, pp. 327-329 (1976).

108. Messinger, M., and H., Shooman, "Technique for optimum spares allocation: a tutorial review," *IEEE Transactions on Reliability,* Vol. R-19, pp. 156-166 (1970).

109. Misra, K. B., "A simple approach for constrained redundancy optimization problems," *IEEE Transactions on Reliability,* Vol. R-21, No. 1, pp. 30-34 (1972).

110. Misra, K. B., "Reliability optimization of a series-parallel
     system," *IEEE Transactions on Reliability,* Vol. R-21, No. 4,
     pp. 230-238 (1972).

111. Misra, K. B., and M. D. Ljubojevic, "Optimal reliability design
     of a system: a new look," *IEEE Transactions on Reliability,* Vol.
     R-22, pp. 255-258 (1973).

112. Misra, K. B., "A method of solving redundancy optimization pro-
     blems," *IEEE Transactions on Reliability,* Vol. R-20, No. 3,
     pp. 117-120 (1971).

113. Misra, K. B., "On optimal reliability design:  A review," *IFAC
     for 6th World Conference,* Boston, Mass. pp. 3.4,1-3.4,10 (1975).

114. Misra, K. B., and J. Sharma, " A new geometric programming for-
     mulation for a reliability problem," *International Journal of
     Quality Control,* Vol. 18, No. 3, pp. 497-503 (1973).

115. Misra, K. B., and C. E. Carter, "Redundancy allocation in a sys-
     tem with many stages," *Microelectronics and Reliability,* Vol.
     12, pp. 223-228 (1973).

116. Misra, K. B., and J. Sharma, "Reliability optimization of a
     system by zero-one programming," *Microelectronics and Reliability,*
     Vol.  12, pp. 229-233 (1973).

117. Misra, K. B., "A method for redundancy allocation," *Microelec-
     tronics and Reliability,* Vol. 12, pp. 389-393 (1973).

118. Misra, K. B., "Reliability optimization through sequential sim-
     plex search," *Int. J. Control,* Vol. 18, pp. 173-183 (1973).

119. Misra, K. B., and J. Sharma, "Reliability optimization with
     integer constraint coefficients," *Microelectronics and Reliability,*
     Vol. 12, pp. 431-433 (1973).

120. Mizukami, K., "Optimum redundancy for maximum system reliability
     by the method of convex and integer programming," *Operations
     Research,* Vol. 16, pp. 392-406 (1968).

121. Morrison, D. F., "The optimum allocation of spare components
     in system," *Technometrics,* Vol. 3, No. 3, pp. 399-406 (1961).

122. Moskowitz, F., and J. B. McLean, "Some reliability aspects of
     system design," *IRE Transactions on Reliability and Quality
     Control,* Vol. PGRQC-8, pp. 7-35, (1956).

123. Myers, B. L., and N. L. Enrich, "Algorithmic optimization of
     system reliability," Annual Technical Conference Transactions,
     American Society for Quality Control, pp. 455-460 (1968).

124. Nelson, A. C., Jr. I. R. Batts and R. L. Beadles, "A computer
     program for approximating system reliability," *IEEE Transactions
     on Reliability,* Vol. R-19, pp. 61-60 (1970).

125. Neuner, G. E., and R. N. Miller, "Resource allocation for max-
     imum reliability," Proceedings of 1966 Annual Symposium on Reli-
     ability, pp. 332-346 (1966).

126. Proschan, F., "Redundancy for reliability improvement," in *Sta-
     tistical Theory of Reliability,* (ed. M. Zelen), Madison: The
     University of Wisconsin Press, pp. 55-70 (1964).

127. Proschan, F., and T. A. Bray, "Optimum redundancy under multiple
     constraints," Operations Research, Vol. 13, pp. 800-814 (1965).

128. Rudd, D. F., "Reliability theory in chemical system design,"
     *I&EC Fundamental,* Vol. 1, No. 2, pp. 138-143 (1962).

129. Sasaki, M., "A simplified method of obtaining highest system
     reliability," *Proceedings of the Eighth National Symposium on
     Reliability and Quality Control,* pp. 489-502 (1962).

130. Sasaki, M., "An easy allotment method achieving maximum system
     reliability," *Proceedings of the Ninth National Symposium on
     Reliability and Quality Control,* pp. 109-124 (1963).

131. Selman, V., and N. T. Grisamore, "Optimum system analysis by
     linear programming," *1966 Processing Annual Symposium on Reli-
     ability,* American Society for Quality Control, pp. 696-703 (1966).

132. Sharma, J., and K. V. Venkateswaran, "A direct method for maxi-
     mizing the system reliability," *IEEE Transactions on Reliability,*
     Vol. R-20, No. 4, pp. 256-259 (1971).

133. Shershin, A. C., "Mathematical optimization techniques for the
     simulataneous apportionments of reliability and maintainability,"
     *Operations Research,* Vol. 18, No. 1, pp 95-106 (1970).

134. Shetty, H. V. K., and D. P. Sengupta, "Reliability optimization
     using SLUMT," *IEEE Transactions on Reliability,* Vol. R-24, No. 1,
     pp. 80-81 (1975).

135. Tillman, F. A., "Integer programming solutions to constrained
     reliability optimization problems," *Transactions of Twentieth
     Annual Technical Conference,* American Society for Quality Con-
     trol, paper number 66-174, pp. 676-693 (1966).

136. Tillman, F. A., C. H. Lie, and C. L. Hwang, "Analysis of pseudo-
     reliability of a combat tank system and its optimal design,"
     *IEEE Transactions on Reliability,* Vol. R-25, pp. 239-242 (1976).

137. Tillman, F. A., and J. M. Liittschwager, "Integer programming formulation of constrained reliability problems," *Management Science*, Vol. 13, No. 11, pp. 887-899 (1967).

138. Tillman, F. A., C. L. Hwang, L. T. Fan, and S. A. Balbale, "Systems reliability subject to multiple nonlinear contraints," *IEEE Transactions on Reliability*, Vol. R-17, No. 3, pp. 153-157 (1968).

139. Tillman, F. A., "Optimization by integer programming of constrained reliability problems with several modes of failure," *IEEE Transactions on Reliability*, Vol. R-18, No. 2, pp. 47-53 (1969).

140. Tillman, F. A., C. L. Hwang, L. T. Fan, and K. C. Lai, "Optimal reliability of a complex system," *IEEE Transactions on Reliability*, Vol. R-19, No. 3, pp. 95-100 (1970).

141. Webster, L. R., "Optimum system reliability and cost effectiveness," *Proceedings of 1967 Annual Symposium on Reliability*, pp. 489-500 (1967).

142. Webster, L. R., "Choosing optimum system configurations," *Proceedings of the Tenth National Symposium on Reliability & Quality Control*, Washington, D. C., pp. 345-359 (1964).

143. Woodhouse, C. F., "Optimal redundancy allocation by dynamic programming," *IEEE Transactions on Reliability*, Vol. R-21, No. 1, pp. 60-62 (1972).

144. Lie, C. H., C. L. Hwang, and F. A. Tillman, "Availability of maintained systems:  a state-of-the-art survey," *AIIE Transactions*, Vol. 9, pp. 247-259 (1977).

## 3.1 INTRODUCTION

It is well-known that by using redundancy we can increase system
reliability.  Many techniques have been applied to obtain the solution
of optimization problems; however, several heuristic approaches
are very attractive for solving the redundancy allocation problems.

Four heuristic approaches are presented in this chapter. Sharma
and Venkateswaran [6] developed an intuitive procedure for allocating
redundancy among subsystems.  The procedure is to add a redundancy
at each iteration in the stage which has the lowest reliability.
The algorithm was used to solve multistage system problems which
were subject to multiple nonlinear constraints.  In this approach the
constraints are never active.  Misra [4] introduced an approach for
solving a redundancy optimization problem with multiple linear constraints.  In the solution process, the problem with r-constraints is
decoupled into r-problems, each having one constraint.  A "desirability
factor", i.e., the ratio of the percentage increase in the system reliability to the percentage increase of the corresponding cost, is
introduced to determine the stage to which a redundancy is to be
added. Aggarwal et al. [2] improved the Sharma and Venkateswaran
approach for solving series system problems with multiple nonlinear
constraints by introducing a relative increment in reliability
versus decrement in the slack variables (the unused balance of
the resources) as a criterion in selecting the stage to which a
redundancy is to be added.  Aggarwal [1] then extended the approach

to the problem of complex systems.  Recently, Nakagawa and Nakashima
[5] presented the fourth approach in solving another type of series
system (described later).  In this approach the balance between
the objective function and the constraints is considered.  An extended
Nakagawa and Nakashima approach for solving complex system problems
is also presented.

These approaches are illustrated by four examples.  Each of the
methods are applied to one or more of the examples.  The first
example is a five-stage series system with three nonlinear constraints.
The second is a complex (nonseries-nonparallel) system to which
one linear constraint is imposed.  The third is a four-stage series
system with two linear cost constraints.  The fourth is a more complex
one which cannot be reduced to a simple parallel or series redundancy
problem. These examples are presented here.

*Example 3-1.*  This problem was presented originally by Tillman and
Liittschwager [7], and used by Tillman et al. [8,9], Sharma and Venkateswaran [6], and others for demonstrating a number of optimization
techniques.

The five-stage problem is stated as
Maximize

$$R_s = \prod_{j=1}^{5} \left[ 1 - (1 - R_j)^{x_j} \right] \tag{1}$$

subject to

$$g_1 = \sum_{j=1}^{5} p_j (x_j)^2 \leq P$$

$$g_2 = \sum_{j=1}^{5} c_j (x_j + \exp (x_j/4)) \leq C$$

$$g_3 = \sum_{j=1}^{5} w_j x_j \exp (x_j/4) \leq W \tag{2}$$

where $x_j \geq 1$, $j = 1,2,\ldots,5$ are integers.

The constants associated with the five-stage problem are:

| $j$ | $R_j$ | $P_i$ | $p$ | $C_j$ | $C$ | $w_j$ | $W$ |
|---|---|---|---|---|---|---|---|
| 1 | 0.80 | 1 | | 7 | | 7 | |
| 2 | 0.85 | 2 | | 7 | | 8 | |
| 3 | 0.90 | 3 | 100 | 5 | 175 | 8 | 200 |
| 4 | 0.65 | 4 | | 9 | | 6 | |
| 5 | 0.75 | 2 | | 4 | | 9 | |

*Example 3-2.* A second example is considered where we have a nonseries-non-parallel system shown in Fig. 1 [1]. There is one cost constraint and a maximum of 20 units to be considered. The data for the various subsystems are

$$R_1 = 0.70, \ R_2 = 0.85, \ R_3 = 0.75, \ R_4 = 0.80, \ R_5 = 0.90$$

$$C_1 = 2, \ C_2 = 3, \ C_3 = 2, \ C_4 = 3, \ C_5 = 1$$

The problem is

Minimize

$$Q_s = Q_1'Q_3' + Q_2'Q_4' + Q_1'Q_4'Q_5' + Q_2'Q_3'Q_5' - Q_1'Q_2'Q_3'Q_4'$$

$$- Q_1'Q_3'Q_4'Q_5' - Q_1'Q_2'Q_3'Q_5' - Q_1'Q_2'Q_4'Q_5' - Q_2'Q_3'Q_4'Q_5'$$

$$+ 2Q_1'Q_2'Q_3'Q_4'Q_5' \tag{3}$$

subject to

$$g = \sum_{j=1}^{5} c_j x_j \leq C \tag{4}$$

where $Q_j' = (1 - R_j)^{x_j}$ , $x_j \geq 1$, $j = 1, 2, \ldots, 5$ are integers.

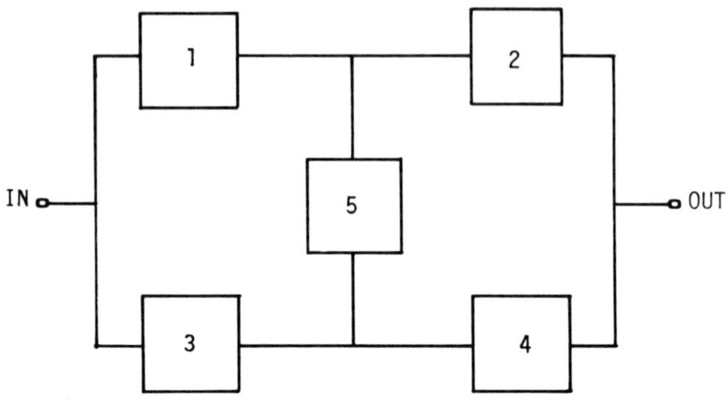

Fig. 1    A bridge configuration.

*Example 3-3.*    The next example is a series system with four stages.
The component reliability, cost, and weight data are:

| Stage, $j$ | 1 | 2 | 3 | 4 |
|---|---|---|---|---|
| Component reliability, $R_j$ | 0.80 | 0.70 | 0.75 | 0.85 |
| Cost, $c_j$ | 1.2 | 2.3 | 3.4 | 4.5 |
| Weight, $w_j$ | 5 | 4 | 8 | 7 |

The system cost and weight are 56 and 120, respectively.

The problem is

Maximize

$$R_s = \prod_{j=1}^{4} [1 - (1 - R_j)^{x_j}]$$

subject to

$$g_1 = \sum_{j=1}^{4} c_j x_j \leq 56$$

$$g_2 = \sum_{j=1}^{4} w_j x_j \leq 120$$

where $x_j \geq 1$, $j = 1,2,3,4$ are integers.

*Example 3-4.*  This example was presented by Nakagawa and Nakashima
[5] and is more complex than a simple parallel-series redundancy problem.

Consider a system composed of 3 stages operating in series. The
system reliability is increased by choosing the most reliable component
out of 4 candidates at stage 1, adding redundant components in parallel
at stage 2 and using 2-out-of-$(x_3 + 1)$:  G configuration at stage
3.[†] The problem is

Maximize

$$R_s = \prod_{j=1}^{3} R_j'(x_j) \tag{5}$$

subject to

$$g_1 = 4 * \exp\left\{ \frac{0.02}{1-R_1(x_1)} \right\} + 5x_2 + 2(x_3 + 1) \leq 45$$

$$g_2 = 5 + e^{x_1/8} + 3(x_2 + e^{x_2/4}) + 5(x_3 + 1 + e^{x_3/4}) \leq 75 \tag{6}$$

$$g_3 = 10 + 8x_2 * e^{x_2/4} + 6x_3 * e^{x_3/4} \leq 240$$

where

$$R_1'(x_1) = 0.88, \ 0.92, \ 098, \ 0.99 \text{ for } x_1 = 1,2,3,4 \text{ respectively}$$

$$R_2'(x_2) = 1 - (1 - 0.81)^{x_2}$$

$$R_3'(x_3) = \sum_{k=2}^{x_3+1} \binom{x_3 + 1}{k} * 0.77^k (1 - 0.77)^{x_3+1-k}$$

---

[†] k-out-of-n:  G  The system is good if and only if at least k out of
its n elements are good.
k-out-of-n:  F  The system is failed if and only if at least k out of
its n elements are failed.

*Example 3-5.*  A multi-function system is considered in this example which contains N distinct components and the reliability of each is increased by adding redundancies in parallel or standby.  The problem is stated by Ushakov [10] as

Minimize

$$C_s = \sum_{j=1}^{N} c_j x_j$$

subject to

$$R_{si} = \prod_{j \in J_i} R_j'(x_j) \geq R_{si,min} \qquad i = 1,2,\ldots,k$$

$$x_j \geq 0, \; j = 1,2,\ldots,N \text{ are integers}$$

where $J_i$ denotes the subset of components that ensures the execution of function i with the minimum requirement for the $i^{th}$ functional reliability, $R_{si}$, $i = 1,2,\ldots,k$.

The data associated with this example are:

| Stage, j | 1 | 2 | 3 | |
|---|---|---|---|---|
| Component reliability, $R_j$ | 0.8 | 0.75 | 0.85 | $R_{s1,min} = 0.94$ |
| Cost, $c_j$ | 3 | 2 | 4 | $r_{s2,min} = 0.96$ |

## 3.2    A HEURISTIC METHOD:  SHARMA AND VENKATESWARAN'S APPROACH

### Formulation of the Problem

In addition to the general assumptions made for the system reliability optimization problems of an N-stage in series with $x_j$ redundant components at stage j, the unreliability of one component at the $j^{th}$ stage, $Q_j$, $j = 1,2,\ldots,N$ should be small enough ($\leq 0.5$)

so that

$$Q_s = 1 - \prod_{j=1}^{N} (1 - Q_j^{x_j})$$

can be approximated by

$$Q_s \simeq \sum_{j=1}^{N} Q_j^{x_j}$$

where $Q_s$ is the system unreliability.  Therefore, the system reliability problem subject to nonlinear cost constraints can be formulated as

Minimize

$$Q_s = \sum_{j=1}^{N} Q_j^{x_j} \tag{7}$$

subject to

$$\sum_{j=1}^{N} g_{ij}(x_j) \le b_i \qquad i = 1,2,\ldots,4 \tag{8}$$

where $g_{ij}(x_j)$ is the resource i consumed in stage j, and $b_i$ is the available resource for constraint i.

The objective is to reduce $Q_s$ in successive steps.  The procedure at each step is to add one redundant component to the stage with the highest $Q_j^{x_j}$ in eq. (7), if constraints in eq. (8) are not violated.  Therefore, the constraints become active only in the neighborhood of the boundary of the feasible region.  The sequential steps involved in solving the problem are as follows:

Step 1.  Assign $x_j = 1$ for $j = 1,2,\ldots,N$.  Because this is a cascade system, there must be at least one component in each stage and the system should not violate any constraints.

Step 2.  Find the stage which is the most unreliable.  Add a redundant component to that stage.

Step 3.  Check the constraints:

   (a)  If any constraint is violated, go to Step 4.

    b)   If no constraint has been violated, go to Step 2.

    c)   If any constraint is exactly satisfied, stop.  The current $x_j$'s are then the optimum configuration of the system.

Step 4.  Remove the redundant component added in Step 2.  The resulting number is the optimum allocation for that stage.  Remove this stage from further consideration.

Step 5.  If all stages have been removed from consideration, the current $x_j$'s are the optimum configuration of the system.  Otherwise, go to Step 2.

*Solution of Example 3-1.*  A five-stage problem with three nonlinear constraints is solved.  The objective of the problem, eq. (1) can be approximated by

Minimize

$$Q_s = (1 - R_1)^{x_1} + (1 - R_2)^{x_2} + (1 - R_3)^{x_3} + (1 - R_4)^{x_4} + (1 - R_5)^{x_5}$$

where $(1 - R_1)^{x_1}$, $(1 - R_2)^{x_2}$, $(1 - R_3)^{x_3}$, $(1 - R_4)^{x_4}$, $(1 - R_5)^{x_5}$ are stage unreliabilities and are represented by $Q_1'$, $Q_2'$, $Q_3'$, $Q_4'$, $Q_5'$, respectively.

The basic allocation (1, 1, 1, 1, 1) is assigned to the system. The stage unreliabilities under this configuration are (0.20, 0.15, 0.10, 0.35, 0.25).  The resources consumed are (12, 73.1, 48.8) and no constraints are violated.  Since stage 4 is the most unreliable, i.e., $Q_4' = 0.35$, we add one redundancy to this stage to form the system configuration (1, 1, 1, 2, 1) and consume resources equal to (24, 85.4, 60.8) respectively.  Following the steps of the algorithm, we obtain the results presented in Table 1.  The optimum result is $(x_1, x_2, x_3, x_4, x_5) = (3, 2, 2, 3, 3)$ with the system reliability of $R_s \simeq 1 - (0.008 + 0.0225 + 0.01 + 0.04288 + 0.01562) = 0.90$.

It is worth noting that with the optimal configuration (3, 2, 2, 3, 3), no constraints are violated, $Q_4'$ is removed from further consideration, and $Q_2' = 0.0225$ is the highest unreliability, so a redundant component may be added to stage 2 to form a new system configuration (3, 3, 2, 3, 3) [Step 2].  However, constraint 3 is

Table 1    Results of Example 3-1 by Sharma and Venkateswarn's Method

| Number of Components in Stage | | | | | Stage Unreliability | | | | | Constraints | | |
|---|---|---|---|---|---|---|---|---|---|---|---|---|
| $x_1$ | $x_2$ | $x_3$ | $x_4$ | $x_5$ | $Q_1'$ | $Q_2'$ | $Q_3'$ | $Q_4'$ | $Q_5'$ | $g_1(\bar{x})$ | $g_2(\bar{x})$ | $g_3(\bar{x})$ |
| 1 | 1 | 1 | 1 | 1 | 0.2 | 0.15 | 0.1 | $0.35^a$ | 0.25 | 12 | 73.1 | 48.8 |
| 1 | 1 | 1 | 2 | 1 | 0.2 | 0.15 | 0.1 | 0.1225 | $0.25^a$ | 24 | 85.4 | 60.8 |
| 1 | 1 | 1 | 2 | 2 | $0.2^a$ | 0.15 | 0.1 | 0.1225 | 0.0625 | 30 | 90.8 | 79.0 |
| 2 | 1 | 1 | 2 | 2 | 0.04 | $0.15^a$ | 0.1 | 0.1225 | 0.0625 | 33 | 100.4 | 93.1 |
| 2 | 2 | 1 | 2 | 2 | 0.04 | 0.0225 | 0.1 | $0.1225^a$ | 0.0625 | 39 | 109.9 | 109.2 |
| 2 | 2 | 1 | 3 | 2 | 0.04 | 0.0225 | $0.1^a$ | 0.042875 | 0.0625 | 59 | 123.12 | 127.5 |
| 2 | 2 | 2 | 3 | 2 | 0.04 | 0.0225 | 0.01 | 0.042875 | $0.0625^a$ | 68 | 130.0 | 143.6 |
| 2 | 2 | 2 | 3 | 3 | $0.04^a$ | 0.0225 | 0.01 | $0.042875^b$ | 0.015625 | 78 | 136.0 | 171.1 |
| 3 | 2 | 2 | 3 | 3 | 0.008 | 0.0225 | 0.01 | $0.042875^b$ | 0.015625 | 83 | 146.1 | 192.5 |
| 3 | 3 | 2 | 3 | 3 | | | | | | 93 | 157.4 | $216.9^c$ |
| 3 | 2 | 2 | 3 | 4 | | | | | | 88 | 142.0 | $220.1^c$ |
| 3 | 2 | 3 | 3 | 3 | | | | | | 91 | 153.0 | $208.6^c$ |
| 4 | 2 | 2 | 3 | 3 | | | | | | 88 | 156.2 | $213.9^c$ |
| 3 | 2 | 2 | 4 | 3 | | | | | | 93 | 159.3 | $210.2^c$ |

[a] This is the stage to which a redundant component is to be added.

[b] This indicates that the stage has been removed from further consideration.

[c] The constraint is violated.

violated [Step 3a]; therefore, $x_2 = 2$ is the optimal number and stage
2 is removed from further consideration [Step 4].  Go back to the
(3, 2, 2, 3, 3) configuration. Continuing with the algorithm, a redundancy is tested at stage 5 ($Q_5' = 0.015625$, the largest unreliability
among $Q_1'$, $Q_3'$, and $Q_5'$) [Step 2]; however, constraint 3 is again violated
[Step 3a], therefore, stage 5 is removed from further consideration
[Step 4].  Similarily, we test stage 3 and stage 1, but constraint
3 is violated in both cases as shown in Table 1.  Therefore, the optimum
configuration for the system is (3, 2, 2, 3, 3).


*Solution of Example 3-2.*  A solution for a complex (nonseries-nonparallel) system is presented.

The basic allocation (1, 1, 1, 1, 1) is again assigned at each
stage for this system.  The stage unreliabilities under this configuration are (0.30, 0.15, 0.25, 0.20, 0.10) which consumes 11 cost
units and obviously does not exceed the available resources.  Since
stage 1 has the highest unreliability, i.e., $Q' = 0.30$, we add
one redundancy to this stage to form the new configuration (2, 1, 1,
1, 1) and check the resource limits.  Using the algorithm, we obtain
the results which are summarized in Table 2.  The last row of Table
2 shows the resource 20 units are consumed with the allocation
of redundancies of (2, 1, 2, 2, 3).  The system unreliability after
substituting (2, 1, 2, 2,3) into eq. (3) is 0.0116, hence the system
reliability is 0.9884.

*Solution of Example 3-3.*  The procedures to reach the solution for
this example are presented in Table 3.


*Solution of Example 3-4.*  The results of this example are listed in
Table 4.

Table 2    Results of Example 3-2 by Sharma and Venkateswaran's Method

| Number of Components in stage | | | | | Stage Unreliabilities | | | | | Constraint |
|---|---|---|---|---|---|---|---|---|---|---|
| $x_1$ | $x_2$ | $x_3$ | $x_4$ | $x_5$ | $Q_1'$ | $Q_2'$ | $Q_3'$ | $Q_4'$ | $Q_5'$ | $g(\bar{x})$ |
| 1 | 1 | 1 | 1 | 1 | 0.30[a] | 0.15 | 0.25 | 0.20 | 0.10 | 11 |
| 2 | 1 | 1 | 1 | 1 | 0.09 | 0.15 | 0.25[a] | 0.20 | 0.10 | 13 |
| 2 | 1 | 2 | 1 | 1 | 0.09 | 0.15 | 0.0625 | 0.20[a] | 0.10 | 15 |
| 2 | 1 | 2 | 2 | 1 | 0.09 | 0.15[b] | 0.0625 | 0.04 | 0.10 | 18 |
| 2 | 2 | 2 | 2 | 1 | | | | | | 21[c] |
| 2 | 1 | 2 | 2 | 2 | 0.09[b] | 0.15[b] | 0.0625[b] | 0.04[b] | 0.01[a] | 19 |
| 2 | 1 | 2 | 2 | 3 | 0.09 | 0.15 | 0.0625 | 0.04 | 0.001 | 20 |

[a]This is the stage to which a redundant component is to be added.

[b]This indicates that the stage has been removed from further consideration.

[c]The constraint is violated.

Table 3    Results of Example 3-3 by Sharma and Venkateswaran's Method

| Number of Components in Stage | | | | Stage Unreliability | | | | Constraints | |
|---|---|---|---|---|---|---|---|---|---|
| $x_1$ | $x_2$ | $x_3$ | $x_4$ | $Q_1'$ | $Q_2'$ | $Q_3'$ | $Q_4'$ | $g_1(\bar{x})$ | $g_2(\bar{x})$ |
| 1 | 1 | 1 | 1 | 0.20 | 0.30[a] | 0.25 | 0.15 | 11.4 | 24 |
| 1 | 2 | 1 | 1 | 0.20 | 0.09 | 0.25[a] | 0.15 | 13.7 | 28 |
| 1 | 2 | 2 | 1 | 0.20[a] | 0.09 | 0.0625 | 0.15 | 17.1 | 36 |
| 2 | 2 | 2 | 1 | 0.04 | 0.09 | 0.0625 | 0.15[a] | 18.3 | 41 |
| 2 | 2 | 2 | 2 | 0.04 | 0.09[a] | 0.0625 | 0.0225 | 22.8 | 48 |
| . | | . | . | . | . | . | . | . | . |
| | | . | . | . | . | . | . | . | . |
| | | . | . | . | . | . | . | . | . |
| 4 | 5 | 5 | 4 | 0.0016 | 0.00243[a] | 0.0009765 | 0.00050625 | 51.3 | 108 |
| 4 | 6 | 5 | 4 | 0.0016[a] | 0.000729 | 0.0009765 | 0.00050625 | 53.6 | 112 |
| 5 | 6 | 5 | 4 | 0.00032 | 0.000729 | 0.0009765 | 0.00050625 | 54.8 | 117 |

[a]This is the stage to which a redundant component is to be added.

Table 4   Results of Example 3-4 by Sharma and Venkateswaran's Method

| Number of Components in Stage | | | Stage Unreliability | | | Constraints | | |
|---|---|---|---|---|---|---|---|---|
| $x_1$ | $x_2$ | $x_3$ | $Q_1'$ | $Q_2'$ | $Q_3'$ | $g_1(\bar{x})$ | $g_2(\bar{x})$ | $g_3(\bar{x})$ |
| 1 | 1 | 1 | 0.12 | 0.19 | $0.4071^a$ | 13.73 | 26.92 | 27.98 |
| 1 | 1 | 2 | 0.12 | $0.19^a$ | 0.13437 | 15.73 | 33.74 | 40.06 |
| 1 | 2 | 2 | 0.12 | 0.0361 | $0.13437^a$ | 20.73 | 40.93 | 56.16 |
| 1 | 2 | 3 | $0.12^a$ | 0.0361 | 0.04027 | 22.73 | 48.27 | 74.48 |
| 2 | 2 | 3 | $0.08^a$ | 0.0361 | 0.04027 | 23.13 | 48.42 | 74.48 |
| 3 | 2 | 3 | 0.02 | 0.0361 | $0.04027^a$ | 28.87 | 48.59 | 74.48 |
| 3 | 2 | 4 | 0.02 | $0.0361^a$ | 0.0114 | 30.87 | 56.69 | 101.61 |
| 3 | 3 | 4 | $0.02^b$ | 0.0069 | 0.0114 | 35.87 | 60.63 | 126.04 |
| 4 | 3 | 4 | | | | $54.55^c$ | 60.82 | 126.04 |
| 3 | 3 | 5 | $0.02^b$ | 0.0069 | 0.0114 | 37.88 | 70.32 | 165.51 |
| 3 | 4 | 5 | | | | 42.88 | $76.39^c$ | $201.68^c$ |
| 3 | 3 | 6 | | | | 39.88 | $78.91^c$ | $222.14^c$ |

[a] This is the stage to which a redundant component is to be added.

[b] This indicates that the stage has been removed from further consideration.

[c] The constraint is violated.

## 3.3  A HEURISTIC METHOD:  AGGARWAL'S APPROACH

### Formulation of the Problem

Sharma and Venkateswaran's heuristic approach consists of adding
redundancies to the stage where the unreliability is the highest.
The method is applicable to problems with any number of general
constraints.  The method may not yield an optimum solution if the stages
have components of similar reliability but quite different cost.
Aggarwal et al.  [2] have proposed an alternate algorithm using
a new criterion for selecting the stage where the redundancy is to
be added.  In certain cases, using Sharma and Venkataswaran's approach,
slack variables (balance of unused resources) prevent the addition
of only one component to the particular stage having the lowest reliability but permit the addition of more than one component to another
stage having a higher reliability. The net increase in reliability
for the latter might be more than that for the former.

This heuristic approach is based on the concept that a component
is added to the stage where its addition produces the greatest ratio
of "increment increases in reliability" to the "product of increments
increase in resources usage."  This ratio is defined by

$$F_j(x_j) = \frac{\Delta(1 - R_j)^{x_j}}{\prod_{i=1}^{3} \Delta g_{ij}(x_j)} \tag{9}$$

where

$$\Delta(1 - R_j)^{x_j} = (1 - R_j)^{x_j} - (1 - R_j)^{x_j+1} = R_j(1 - R_j)^{x_j}$$

and

$$\Delta g_{ij}(x_j) = g_{ij}(x_j + 1) - g_{ij}(x_j)$$

$F_j(x_j)$ is a function of j as well as $x_j$; hence in the computation,
it keeps changing even for a fixed j.  In the case of linear constraints, however, all $F_j(x_j)$ can be evaluated by using the recursive

relationship

$$F_j(x_j + 1) = Q_j F_j(x_j) \tag{10}$$

The computational procedure is:

Step 1.  Let $\bar{x} = (x_1, x_2, \ldots, x_N) = (1, 1, \ldots, 1)$.

Step 2.  a)  Calculate $F_j(x_j)$ for all $j$ using eq. (9).

       b)  Select the stage having the highest $F_j(x_j)$.  A redundant component is proposed to be added to that stage.

Step 3.  Check to see if the constraints are violated.

       a)  If the solution is still feasible, add one redundant component to the stage having the highest $F_j(x_j)$. Modify the value of $x_j$ and hence $F_j(x_j)$ and go to Step 2.

       b)  If at least one constraint is exactly satisfied; the current value of $\bar{x}$ is an optimal solution.

       c)  If at least one constraint is violated, cancel the proposed addition of the redundant component; remove that stage from further consideration and repeat step 2. When all the stages are excluded from further consideration, the current values of $\bar{x}$ are the optimal configuration.

Step 4.  Calculate the system reliability, $R_s$, for the optimum $\bar{x}^*$.

*Solution of Example 3-1.*  To obtain the selection factors, $\Delta g_{ij}(x_j)$, $i = 1, 2, 3, \ldots$ are:

$$\Delta g_{1j}(x_j) = p_j(x_j + 1)^2 - p_j(x_j)^2$$

$$\Delta g_{2j}(x_j) = c_j\{(x_j + 1) + \exp[(x_j + 1)/4]\} - c_j[x_j + \exp(x_j/4)]$$

$$\Delta g_{3j}(x_j) = w_j(x_j + 1)\exp[(x_j + 1)/4] - w_j x_j \exp(x_j/4)$$

and the selection factors are:

$$F_j(x_j) = \frac{R_j Q_j^{x_j}}{\prod\limits_{j=1}^{3} \Delta g_{ij}(x_j)} , \qquad j = 1,2,\ldots,5$$

Starting with $\bar{x} = (1, 1, 1, 1, 1)$ we add one component to a stage at a time as shown in Table 5. The stage selection factors for components $(1, 1, 1, 1, 1)$ are $(0.000396, 0.000138, 0.000091, 0.000128, 0.000316)$. The resources consumed are $(12, 73.09, 49.79)$ which have not violated the constraints. Since stage 1 has the highest stage selection factor i.e., $F_1(x_1) = 0.000396$, we add one redundancy to this stage to form the system configuration $(2, 1, 1, 1, 1)$, and check the consumed resources $(15, 82.64, 62.88)$. Following the steps of the computational procedures, the optimum configuration is $(x_1, x_2, x_3, x_4, x_5) = (3, 2, 2, 3, 3)$. The optimum system reliability is 0.9045.

*Solution of Example 3-2.* This example, solved by Aggarwal's approach, shows that the Sharma and Venkateswaran's approach will not always give an optimum solution.

The initial configuration of each subsystem is $\bar{x} = (1, 1, 1, 1, 1)$. The stage selection factors are $(0.02649, 0.01783, 0.02759, 0.01068, 0.00533)$ and the resources consumed are 11 units. Since subsystem 3 has the highest selection factor, i.e., $F_3(x_3) = 0.02759$, we add one redundancy to this stage to form the system configuration $(1, 1, 2, 1, 1)$ which consumes 13 units of cost. Completing the computational procedures, the optimum system configuration is $(x_1, x_2, x_3, x_4, x_5) = (3, 1, 2, 2, 1)$ as shown in Table 6 with the optimum system reliability of 0.9914. It is noted that this same problem solved by Sharma and Venkateswaran's approach yields a system configuration of $(2, 1, 2, 2, 3)$ and a system reliability of 0.9884 which is not the optimal solution.

Table 5   Results of Example 3-1 by Aggarwal's Method

| Number of components in stage | | | | | Stage selection factor | | | | | Constraints | | |
| --- | --- | --- | --- | --- | --- | --- | --- | --- | --- | --- | --- | --- |
| $x_1$ | $x_2$ | $x_3$ | $x_4$ | $x_5$ | $F_1(x_1)$ | $F_2(x_2)$ | $F_3(x_3)$ | $F_4(x_4)$ | $F_5(x_5)$ | $g_1(\bar{x})$ | $g_2(\bar{x})$ | $g_3(\bar{x})$ |
| 1 | 1 | 1 | 1 | 1 | 0.000396[a] | 0.000138 | 0.000091 | 0.000128 | 0.000316 | 12 | 73.09 | 48.79 |
| 2 | 1 | 1 | 1 | 1 | 0.0000291 | 0.000138 | 0.000091 | 0.000128 | 0.000316[a] | 15 | 82.64 | 62.88 |
| 2 | 1 | 1 | 1 | 2 | 0.0000291 | 0.000138[a] | 0.000091 | 0.000128 | 0.0000290 | 21 | 88.10 | 81.00 |
| 2 | 2 | 1 | 1 | 2 | 0.0000291 | 0.0000076 | 0.000091 | 0.000128[a] | 0.0000290 | 27 | 97.65 | 97.11 |
| 2 | 2 | 1 | 2 | 2 | 0.0000291 | 0.0000076 | 0.000091[a] | 0.000016 | 0.0000290 | 39 | 109.93 | 109.19 |
| 2 | 2 | 2 | 2 | 2 | 0.0000291[a] | 0.0000076 | 0.0000033 | 0.000016 | 0.0000290 | 48 | 116.75 | 125.30 |
| 3 | 2 | 2 | 2 | 2 | 0.00000258 | 0.0000076 | 0.0000033 | 0.000016 | 0.0000290[a] | 53 | 127.03 | 146.67 |
| 3 | 2 | 2 | 2 | 3 | 0.00000258 | 0.0000076 | 0.0000033 | 0.000016[a] | 0.0000032 | 63 | 132.90 | 174.15 |
| 3 | 2 | 2 | 3 | 3 | 0.00000258 | 0.0000076[a] | 0.0000033 | 0.0000025 | 0.0000032 | 83 | 146.11 | 192.47 |
| 3 | 3 | 2 | 3 | 3 | | | | | | 93 | 157.39 | 216.90[c] |
| 3 | 2 | 3 | 3 | 3 | | | | | | 91 | 153.00 | 208.60[c] |
| 3 | 2 | 2 | 3 | 4 | | | | | | 80 | 142.00 | 220.10[c] |
| 4 | 2 | 2 | 3 | 3 | | | | | | 88 | 156.20 | 213.90[c] |
| 3 | 2 | 2 | 4 | 3 | | | | | | 93 | 159.30 | 210.20[c] |

[a] This is the stage to which a redundant component is to be added.

[b] This indicates that the stage has been removed from further consideration.

[c] The constraint is violated.

49

Table 6   Results of Example 3-2 by Aggarwal's method

| Number of components in stage | | | | | Stage selection factor | | | | | Constraint |
|---|---|---|---|---|---|---|---|---|---|---|
| $x_1$ | $x_2$ | $x_3$ | $x_4$ | $x_5$ | $F_1(x_1)$ | $F_2(x_2)$ | $F_3(x_3)$ | $F_4(x_4)$ | $F_5(x_5)$ | $g(\bar{x})$ |
| 1 | 1 | 1 | 1 | 1 | 0.02649 | 0.01783 | 0.02759[a] | 0.01068 | 0.00533 | 11 |
| 1 | 1 | 2 | 1 | 1 | 0.00796 | 0.00825 | 0.000690 | 0.01149[a] | 0.00478 | 13 |
| 1 | 1 | 2 | 2 | 1 | 0.01053[a] | 0.00180 | 0.00720 | 0.00230 | 0.00048 | 16 |
| 2 | 1 | 2 | 2 | 1 | 0.00316[a] | 0.00191 | 0.00160 | 0.00180 | 0.00033 | 18 |
| 3 | 1 | 2 | 2 | 1 | 0.000024 | 0.00030 | 0.00039 | 0.00092 | 0.00009 | 20 |

[a]This is the stage to which a redundant component is to be added.

[b]This indicates that the stage has been removed from further consideration.

50

Table 7  Results of Example 3-3 by Aggarwal's method

| Number of components in stage | | | | Stage selection factor | | | | Constraints | |
| --- | --- | --- | --- | --- | --- | --- | --- | --- | --- |
| $x_1$ | $x_2$ | $x_3$ | $x_4$ | $F_1(x_1)$ | $F_2(x_2)$ | $F_3(x_3)$ | $F_4(x_4)$ | $g_1(\bar{x})$ | $g_2(\bar{x})$ |
| 1 | 1 | 1 | 1 | 1.3333[a] | 0.0913 | 0.0662 | 0.0283 | 11.4 | 24 |
| 2 | 1 | 1 | 1 | 0.2666[a] | 0.0913 | 0.0662 | 0.0283 | 12.6 | 29 |
| 3 | 1 | 1 | 1 | 0.0533 | 0.0913[a] | 0.0662 | 0.0283 | 13.8 | 34 |
| 3 | 2 | 1 | 1 | 0.0533 | 0.0274 | 0.0662[a] | 0.0283 | 16.1 | 38 |
| 3 | 2 | 2 | 1 | 0.0533[a] | 0.0274 | 0.0165 | 0.0283 | 19.5 | 46 |
| 4 | 2 | 2 | 1 | 0.0011 | 0.0274 | 0.0165 | 0.0283[a] | 20.7 | 53 |
| . | | | | . | . | . | . | . | . |
| . | | | | . | . | . | . | . | . |
| . | | | | . | . | . | . | . | . |
| 4 | 5 | 5 | 4 | 0.0011[a] | 0.0007 | 0.0010 | 0.0006 | 51.3 | 104 |
| 5 | 5 | 5 | 4 | 0.0002 | 0.0007[a] | 0.0013[b] | 0.0006 | 52.5 | 109 |
| 5 | 6 | 5 | 4 | 0.0002 | 0.0002 | 0.0013 | 0.0006 | 54.8 | 117 |

[a]This is the stage to which a redundant component is to be added.

[b]This indicates that the stage has been removed from further consideration.

*Solution of Example 3-3.*   Since this example contains linear constraints, all $F_j(x_j)$ can be evaluated by using the recursive relationship of eq. (10).   This equation is more convenient to determine the
selection factors in order to decide where to add redundancies.   The
results are summarized and are shown in Table 7.

## 3.4   A HEURISTIC METHOD:   MISRA'S APPROACH

### Formulation of the Problem

This method is used to solve the redundancy problem with multiple
linear constraints.   Basically, the solution to an r-constraint
problem is obtained by successively solving r-unconstrained problems.
At each step, an active constraint is selected, and then using the
maximum gradient (explained later) a better solution is determined.
The computational procedure is presented in the following 5 steps [4].

Step 1.   Within the feasible solution domain, the attainable reliability should be approximately estimated by the allocation of
redundancies at each stage, until some constrained resource is
limiting.

Step 2.   Using this estimate of system reliability, $R_s$, find the
optimum allocation of redundancies with respect to each "cost"
constraint by

$$x_j = \frac{\log (1 - R_s^{a_j})}{\log Q_j} \tag{11}$$

where

$$a_j = \frac{c_j/(\ln Q_j)}{\sum_{j=1}^{N} c_j/(\ln Q_j)}, \qquad j = 1,2,\ldots,N$$

Usually a different system configuration is obtained for each
constraint.

Step 3.   From these configurations choose the highest system reliability as the reference reliability index for comparison.   To

each configuration having lower system reliability, add one component
to the stage which has the highest desirability factor defined by

$$F_j = \frac{\Delta R_s / R_s}{c_j / b_j} \qquad\qquad j = 1, 2, \ldots, N \qquad\qquad (12)$$

Step 4.  Now, each configuration is improved, except the reference
   reliability index.  Then
   (a)  go to Step 3, if none of these configurations gives the same
        system reliability within the domain of feasible solutions
        (the constraints are not violated and the configuration has
        a higher reliability).
   (b) go to Step 3, if all configurations give the same system
       reliability.  Otherwise, go to Step 5(a).
Step 5.
   (a) Stop, this common configuration is the optimum one.
   (b) If a common level of reliability is not available, the
       configuration with the highest reliability will be close
       to the optimum solution.


*Solution of Example 3-3.*  To find a suboptimal system reliability
which does not violate any of the cost or weight constraints, an enumeration method is used (see Table 8).  The system reliability, $R_s$,
is 0.99577 [Step 1].  Using this system reliability in eq. (11), the
optimal allocations with respect to the cost constraint and the weight
constraint are obtained as (5, 5, 4, 3) and (4, 6, 4, 3), respectively
[Step 2].

Now we can construct Table 9 in the following way.  The reliability levels of system configurations of (5, 5, 4, 3) and (4, 6, 4, 3)
are 0.9900 and 0.9904, respectively.  Since 0.9900 is smaller than
0.9904, we then add one more redundancy to a stage of the system
(5, 5, 4, 3). Since the desirability factors, $F$ , for each stage
are (0.0139, 0.0356, 0.0386, 0.0310), and 0.0386 is the largest, one
redundancy is added to stage 3.  The system reliability for (5, 5,
5, 3) is 0.9929 [Step 3].

Table 8    Suboptimal Results of Example 3-3 by an Enumeration Method

| Stage | | | | Constraints | |
|---|---|---|---|---|---|
| 1 | 2 | 3 | 4 | Cost | Weight |
| 1 | 1 | 1 | 1 | 11.4 | 24 |
| 2 | 1 | 1 | 1 | 12.6 | 29 |
| 2 | 2 | 1 | 1 | 14.9 | 33 |
| 2 | 2 | 2 | 1 | 18.3 | 41 |
| 2 | 2 | 2 | 2 | 22.8 | 48 |
| 3 | 2 | 2 | 2 | 24.0 | 53 |
| 3 | 3 | 2 | 2 | 26.3 | 57 |
| 3 | 3 | 3 | 2 | 29.7 | 65 |
| 3 | 3 | 3 | 3 | 34.2 | 72 |
| 4 | 3 | 3 | 3 | 35.4 | 77 |
| 4 | 4 | 3 | 3 | 38.7 | 81 |
| 4 | 4 | 4 | 3 | 41.1 | 89 |
| 4 | 4 | 4 | 4 | 45.6 | 96 |
| 5 | 4 | 4 | 4 | 46.8 | 101 |
| 5 | 5 | 4 | 4 | 49.1 | 105 |
| 5 | 5 | 5 | 4 | 52.5 | 113 |
| 5 | 5 | 5 | 5 | 57.0 | 120 |

$R_s = 0.99577$

Table 9    Results of Example 3-3 by Misra's Method

| | BASED ON COST CONSTRAINT | | | | | | | | | | BASED ON WEIGHT CONSTRAINT | | | | | | | | | |
|---|---|---|---|---|---|---|---|---|---|---|---|---|---|---|---|---|---|---|---|---|
| Iter-ation | $x_1$ | $x_2$ | $x_3$ | $x_4$ | Cost | $R_s$ | $F_1$ | $F_2$ | $F_3$ | $F_4$ | $x_1$ | $x_2$ | $x_3$ | $x_4$ | Weight | $R_s$ | $F_1$ | $F_2$ | $F_3$ | $F_4$ |
| 1 | 5 | 5 | 4 | 3 | 44.6 | .9900 | .0139 | .0356 | .0386[a] | .0310 | 4 | 6 | 4 | 3 | 97 | .9904 | — | — | — | — |
| 2 | 5 | 5 | 5 | 3 | 48.0 | .9929 | — | — | — | — | 4 | 6 | 4 | 3 | 97 | .9904 | .0287 | .027 | .0382 | .0399[a] |
| 3 | 5 | 5 | 5 | 3 | 48.0 | .9929 | .0139 | .0356[a] | .0290 | .0322 | 4 | 6 | 4 | 4 | 104 | .9933 | — | — | — | — |
| 4 | 5 | 6 | 5 | 3 | 50.3 | .9946 | — | — | — | — | 4 | 6 | 4 | 4 | 104 | .9933 | .0287 | .027 | .0382[a] | .0223 |
| 5 | 5 | 6 | 5 | 3 | 50.3 | .9946 | .0139 | .0310 | .0290 | .0322[a] | 4 | 6 | 5 | 4 | 112 | .9962 | — | — | — | — |
| 6 | 5 | 6 | 5 | 4 | 54.8 | .9975 | — | — | — | — | 4 | 6 | 5 | 4 | 112 | .9962 | .0287[a] | .027 | .0243 | .0223 |
| 7 | 5 | 6 | 5 | 4 | 54.8 | .9975 | — | — | — | — | 5 | 6 | 5 | 4 | 117 | .9975 | — | — | — | — |

[a]This indicates the maximum $F_j$, $j = 1,2,3,4.$

This reliability is better than the system reliability of (0.9904) for (4, 6, 4, 3).  (5, 5, 5, 3) and (4, 6, 4, 3) are obviously not giving the same system reliability and if we select a better solution the constraints are not violated [Step 4a].  We, therefore, add a component to one stage of (4, 6, 4, 3).  Since the desirability factors, $F_j$, for each stage are now (0.0287, 0.0270, 0.0382, 0.0399), and 0.0399 is the largest, we add one redundancy to stage 4.  The resulting system reliability for (4, 6, 4, 4) is 0.9933 [Step 3].  This reliability then is compared with 0.9929.  (5, 5, 5, 3) and (4, 6, 4, 4) are still not giving the same system reliability and if a better solution is selected the constraints will not be violated [Step 4a].  We, therefore, add a component to one stage of (5, 5, 5, 3).  Following the iterative procedures, the common allocation (5, 6, 5, 4) is finally obtained.  The consumed cost resource and weight resource are 54.8 and 117.0, respectively.  Since a redundancy added to any stage will exceed the available resources, (5, 6, 5, 4) is the optimal configuration of the system which gives the system reliability 0.9975 [Step 5a].

## 3.5  A HEURISTIC METHOD:  USHAKOV'S APPROACH

### Formulation of the Problem

Suppose that a system consists of N distinct elements and in order to increase the reliability of each element, we can use an arbitrary type of standby.  Let $R_j'(X_j)$ denote the probability of failure-free operation of element j when $(X_j - 1)$ standby elements are used to ensure its operability, let $c_j$ denote the cost of a single element of type j, and let $J_i$ denote the subset of system elements that ensure execution of function i with given probability $R_{si,min}$ $(i = 1, \ldots, k)$. With this notation, we can formulate the following problem [10]:

Minimize

$$\sum_{j=1}^{N} c_j x_j$$

subject to

$$\prod_{j \in J_i} R'_j(X_j) \geq R_{si,min} \qquad i = 1, \ldots, k$$

where $X_j$, $j = 1, \ldots, N$ is a rational number.

The computational procedure can be stated as:

Step 1.  Solve k problems by finding the min $\sum_{j \in J_i} c_j x_j$ under the

condition that

$$\prod_{j \in J_i} R'_j(X_j) \geq R_{si,min} \quad (i = 1, \ldots, k). \quad \text{For problem i, we find}$$

the corresponding optimum values $X_1^i, \ldots, X_N^i$.

Step 2.  For element i, we find the greatest value, that is, $X_j^* = \max X_j^i$.

Step 3.  For each subset of elements $J_i$, we find a smaller subset (which we denote by $J_i^{**}$) that is necessary only for executing function i.  We denote by $J_i^*$ the remaining portion of the subset $J_i$.

Step 4.  For each subset $J_i^*$, we find the value of $\prod_{j \in J_i^*} R'_j(X_j^*) = R_i^*$.

If $J_i^*$ is empty, that is, if function i is executed with an independent group of elements, we take $R_i^* = 1$.

Step 5.  For each set $J_i$, we calculate $R_i^{**} = R_{si,min} / R_i^*$, which is a requirement on the probability of failure-free operation of the elements belonging to subset $J_i^{**}$.

Step 6.  In addition we solve k problems of finding min $\sum\limits_{j \in J_i^*} c_j X_j$

under the condition that $\sum\limits_{j \in J_i^{**}} R_j'(X_j) \geq R_i^{**}$ (i = 1, ..., k).  We

find the values of $X_j^{**}$.

Step 7.  For the solution we take the values of $m_j^*$ for $j \in J_i^*$ found in Step 2, and the values of $X_j^{**}$ for $j \in \bar{J}_i^{**}$ found in Step 6.  Steps 3 through 6 are introduced in order to lower, if possible, the superfluously high reliability in conservation of the given requirements on the probability of execution of the individual functions.

*Solution of Example 3-5.*  Following the computational procedure, we obtain

(1)   $X_1' = 2$                    $X_2' = 3$

     $X_1^2 = 3$                    $X_3^2 = 2$

(2)   $X_1^* = 3$

     $X_2^* = 3$

     $X_3^2 = 2$

(3)   $J_i^{**} = \{1, 2\}$      $J_2^{**} = \{1, 3\}$

     $J_1^* = \{3\}$            $J_2^* = \{2\}$

(4)   $R_1^* = 1 - (1 - 0.85)^2 = 0.9775$

     $R_2^* = 1 - (1 - 0.75)^2 = 0.9844$

(5)　$R_1^{**} = 0.94/0.9775 = 0.9616$

$R_2^{**} = 0.96/0.9844 = 0.9752$

(6)　$X_1^{**} = 3$

$X_2^{**} = 3$

$X_3^{**} = 2$

(7)　Since both the configurations $(X_1^{**}, X_2^{**}, X_3^{**}) = (3, 3, 2)$ and $(X_1, X_2, X_3) = (3, 3, 2)$ satisfy the reliability constraints and the former one costs only 23 which is less than the 26 of the second one, therefore, the optimal solution is $(3, 3, 2)$.

## 3.6　A HEURISTIC METHOD:　NAKAGAWA AND NAKASHIMA'S APPROACH

### Formulation of the Problem

In the previous three approaches, it is assumed that the component unreliability at each stage, $Q_j$, $j = 1, 2, \ldots, n$ is small so that the objective function can be approximated. However, no approximation on the objective function is made in Nakagawa and Nakashima's approach.

We again state the general nonlinear optimization problem (Problem A) for an N-stage series system as [5]:

*Problem A.*　Maximize

$$R_s = \prod_{j=1}^{N} R_j'(x_j)$$

subject to

$$\sum_{j=1}^{N} g_{ij}(x_j) \leq b_i, \qquad i = 1, 2, \ldots, r$$

$$1 \leq x_j \leq \bar{x}_j, \qquad j = 1, 2, \ldots, N$$

where, $\bar{x}_j$ is the maximum number of components used at stage $j$, and all $x_j$'s are integers.

Based on the definitions of

$$\Delta f_j(x_j) = \begin{cases} \ln R'_j(x_j) \, , & \text{for } x_j = 1 \\ \ln R'_j(x_j) - \ln R'_j(x_j - 1), & \text{for } x_j > 1 \end{cases} \qquad (13)$$

and

$$\Delta g_{ij}(x_j) = \begin{cases} g_{ij}(x_j), & \text{for } x_j = 1 \\ g_{ij}(x_j) - g_{ij}(x_j - 1), & \text{for } x_j > 1 \end{cases} \qquad (14)$$

We transform Problem A to Problem B as follows:

$$\ln R_s = \ln \prod_{j=1}^{N} R'_j(x_j)$$

$$= \sum_{j=1}^{N} \ln R'_j(x_j)$$

Based on the definition in eq. (13)

$$\text{if } x_j = 1, \qquad \sum_{j=1}^{N} \ln R'_j(1) = \sum_{j=1}^{N} \Delta f_j(x_j),$$

$$\sum_{j=1}^{N} g_{ij}(1) = \sum_{j=1}^{N} \Delta g_{ij}(x_j)$$

$$\text{if } 1 < x_j \leq \bar{x}_j,$$

$$\sum_{j=1}^{N} \ln R'_j(x_j) = \sum_{j=1}^{N} \left\{ [\ln R'_j(\bar{x}_j) - \ln R'_j(\bar{x}_j - 1)] \right.$$

$$+ [\ln R'_j(\bar{x}_j - 1) - \ln R'_j(\bar{x}_j - 2)] + \ldots$$

$$\left. + [\ln R'(2) - \ln R'(1)] + \ln R'(1) \right\}$$

$$= \sum_{j=1}^{N} \sum_{x_j=1}^{\bar{x}_j} \Delta f_j(x_j)$$

$$\sum_{j=1}^{N} g_{ij}(x_j) = \sum_{j=1}^{N} \{[g_{ij}(\bar{x}_j) - g_{ij}(\bar{x}_j - 1)]$$

$$+ [g_{ij}(\bar{x}_j - 1) - g_{ij}(\bar{x}_j - 2)] + \ldots$$

$$+ [g_{ij}(2) - g_{ij}(1)] + [g_{ij}(1)]\}$$

$$= \sum_{j=1}^{N} \sum_{x_j=1}^{\bar{x}_j} \Delta g_{ij}(x_j)$$

*Problem B.*  Maximize

$$\ln R_s = \sum_{j=1}^{N} \sum_{x_j=1}^{\bar{x}_j} \Delta f_j(x_j)$$

subject to

$$\sum_{j=1}^{N} \sum_{x_j=1}^{\bar{x}_j} \Delta g_{ij}(x_j) \leq b_i, \qquad i = 1,2,\ldots,r$$

$$1 \leq x_j \leq \bar{x}_j, \qquad j = 1,2,\ldots,N$$

where, $\bar{x}_j$ is the maximum number of components for stage j, and all $x_j$'s are integers.

Since $R_j(x_j)$ and $g_{ij}(x_j)$ are monotonically increasing functions in $x_j$ for $j = 1,2,\ldots,N$ and $i = 1,2,\ldots,r$, then

$$\Delta f_j(x_j) \geq 0, \qquad \text{for all } j \text{ and } x_j,$$

and

$$\Delta g_{ij}(x_j) \geq 0, \qquad \text{for all } i, j, \text{ and } x_j.$$

### The Computational Procedure

The computational procedure for solving Problem B is stated as follows.

Step 1.  Set the first solution as

$$x^C = (x_1^C, x_2^C, \ldots, x_N^C) = (1, 1, \ldots, 1)$$

Step 2.  Calculate $b_i^C$ for all i, where

$$b_i^C = b_i - \sum_{j=1}^{N} \sum_{x_j=1}^{x_j^C} \Delta g_{ij}(x_j) \tag{15}$$

Step 3.  Calculate $\Delta x_j$ for all j, where

$$\Delta x_j \equiv \min_{i} \{ b_i^C / \Delta g_{ij}(x_j^C + 1) \}$$

Step 4.  Let $L_{+1} = \{ j \mid \Delta x_j \geq 1 \}$

If $L_{+1}$ is empty, stop.  An optimal solution is obtained.  Otherwise go to Step 5.

Step 5.  Search over m such that

$$S_m = \max_{j \in L_{+1}} \{ S_j \}$$

where

$$S_j \equiv \Delta f_j(x_j^C + 1) \cdot [(1 - \alpha) \cdot \min_{\ell \in L_{+1}} \{ \Delta x_\ell \} + \alpha \Delta x_j] \tag{16}$$

Step 6.  If $x_m^C = \bar{x}_m$, then $x_m^C$ is the optimal number used in stage m, exclude this stage and go to Step 3.  If $x_m^C < \bar{x}_m$, then set $x_m^C = x_m^C + 1$ and go to Step 3.

The above procedure is for a given balancing coefficient "$\alpha$". Nakagawa and Nakashima use these balancing coefficients as an integral part of their solution method.  These $\alpha$'s are simply systematic search values that assist in obtaining an overall optimal solution. One simply selects a range of values for $\alpha$, then evaluates the solution for each of these values. For our example the optimal solutions for the set of "$\alpha$" ($\alpha$ = 0, 0.1, ..., 0.9, 1.0, $1/\alpha$ = 0.9, 0.6, 0.3) should be obtained. The best solution among the solutions for the given set of "$\alpha$'s" is the optimal solution.

*Solution of Example 3-1.* The problem, after transformation, is formed as follows:

Maximize

$$\ln R_s = \sum_{j=1}^{5} \sum_{x_j=1}^{\bar{x}_j} \Delta f_j(x_j)$$

subject to

$$\sum_{j=1}^{5} \sum_{x_j=1}^{\bar{x}_j} \Delta g_{1j}(x_j) \leq P$$

$$\sum_{j=1}^{5} \sum_{x_j=1}^{\bar{x}_j} \Delta g_{2j}(x_j) \leq C$$

$$\sum_{j=1}^{5} \sum_{x_j=1}^{\bar{x}_j} \Delta g_{3j}(x_j) \leq W$$

$$1 \leq x_j \leq \bar{x}_j, \qquad j = 1, 2, \ldots, 5$$

where $x_j$'s are integers, $\Delta f_j(x_j)$ and $\Delta g_{ij}(x_j)$ are given in eqs. (13) and (14).

Set the first solution as $x^c = (x_1^c, x_2^c, x_3^c, x_4^c, x_5^c) = (1, 1, 1, 1, 1)$ [Step 1]. The current resources available $(b_1^c, b_2^c, b_3^c)$ are calculated by eq. (15) and are (98, 101.9, 151.21) [Step 2]. Then the quantities obtained from the constraints at the current solution, $(\Delta x_1, \Delta x_2, \Delta x_3, \Delta x_4, \Delta x_5,)$ are calculated to be (10.67, 9.39, 9.39, 8.17, 8.39) [Step 3]. Since $\Delta x_j$, j = 1,2,3,4,5 are greater than 1, then $L_{+1}$ = [1, 2, 3, 4, 5], which means the redundancy to any stage will not violate the constraints [Step 4]. We can find the stage sensitivities, $(S_1, S_2, S_3, S_4, S_5)$, to be (1.717, 1.227, 0.837, 2.451, 1.847), if $\alpha$ = 0.50. Since stage 4 shows the most sensitive effect, i.e., $S_4$ = 2.451, we should add one redundancy to this stage [Step 5]. Following the steps of the algorithm, we obtain the results as presented in Table 10. The last system configuration in Table 10 is (3, 2, 2, 3, 3) and all the $\Delta x_j$, j = 1,2,3,4,5, are less than 1. This means that it is impossible to add another redundancy to any stage, and (3, 2, 2, 3, 3) is the optimum solution. This is the same solution obtained by Sharma and Venkateswaran's approach and by Aggarwal et al.'s approach.

*Solution of Example 3-4.* This example is presented in Nakagawa and Nakashima's paper. The results are summarized in Table 11. The optimal solution is $(x_1, x_2, x_3)$ = (3, 3, 5). The resources consumed are 37.88, 70.32, and 211.8 for $g_1$, $g_2$, and $g_3$ respectively.

## Extension of Nakagawa and Nakashima's Approach for Complex System Optimization [3]

Nakagawa and Nakashima's approach is used to only solve series-system problems. For a complex system, e.g., Example 3-2, the problem does not have the objective function in the form of Problem A. Therefore, the problem cannot be transformed into Problem B.

To solve the problem of a complex configuration, $\Delta f_j(x_j)$ is redefined as follows:

$$\Delta f_j(x_j) = Q_s(Q_1', \ldots, (Q_j' = Q_j^{x_j}), \ldots, Q_k')$$

$$- Q_s(Q_1', \ldots, (Q_j' = Q_j^{x_j+1}), \ldots, Q_k')$$

$$\simeq \frac{\partial Q_s}{\partial Q_j'} (Q_j^{x_j} - Q_j^{x_j+1})$$

$$= (1 - Q_j) \, Q_j^{x_j} \, \frac{\partial Q_s}{\partial Q_j'}$$

$$= R_j Q_j^{x_j} \, \frac{\partial Q_s}{\partial Q_j'} \tag{17}$$

With this definition of $\Delta f_j(x_j)$ we can then follow the same computational procedures to obtain the optimal solution.

*Solution of Example 3-2.* To solve, we have to use eq. (17) to determine $\Delta f_j(x_j)$. In eq. (17), $\dfrac{\partial Q_s}{\partial Q_j'}$, $j = 1, 2, \ldots, 5$ for this example are given by

$$\frac{\partial Q_s}{\partial Q_1'} = Q_3' + Q_4'Q_5' - Q_2'Q_3'Q_4' - Q_3'Q_4'Q_5' - Q_2'Q_3'Q_5' - Q_2'Q_4'Q_5' + 2\, Q_2'Q_3'Q_4'Q_5'$$

$$\frac{\partial Q_s}{\partial Q_2'} = Q_4' + Q_3'Q_5' - Q_1'Q_3'Q_5' - Q_1'Q_3'Q_4' - Q_1'Q_4'Q_5' - Q_3'Q_4'Q_5' + 2\, Q_1'Q_3'Q_4'Q_5'$$

$$\frac{\partial Q_s}{\partial Q_3'} = Q_1' + Q_2'Q_5' - Q_1'Q_2'Q_4' - Q_1'Q_4'Q_5' - Q_1'Q_2'Q_5' - Q_2'Q_4'Q_5' + 2\, Q_1'Q_2'Q_4'Q_5'$$

$$\frac{\partial Q_s}{\partial Q_4'} = Q_2' + Q_1'Q_5' - Q_1'Q_2'Q_3' - Q_1'Q_3'Q_5' - Q_1'Q_2'Q_3' - Q_2'Q_3'Q_5' + 2\, Q_1'Q_2'Q_3'Q_5'$$

$$\frac{\partial Q_s}{\partial Q_5'} = Q_1'Q_4' + Q_2'Q_3' - Q_1'Q_3'Q_4' - Q_1'Q_2'Q_3' - Q_1'Q_2'Q_4' - Q_2'Q_3'Q_4' + 2\, Q_1'Q_2'Q_3'Q_4'$$

The procedure to obtain the optimal result is presented in Table 12. The optimal result is $(x_1, x_2, x_3, x_4, x_5) = (3, 1, 2, 2, 1)$. The system reliability is 0.9914. The resource, 20, is totally consumed.

Table 10  Results of Example 3-1 by Nakagawa and Nakashima's approach

| Number of components in each stage | | | | | Available resources | | | | | | | | Stage sensitivity | | | | |
|---|---|---|---|---|---|---|---|---|---|---|---|---|---|---|---|---|---|
| $x_1$ | $x_2$ | $x_3$ | $x_4$ | $x_5$ | $b_1^c$ | $b_2^c$ | $b_3^c$ | $\Delta x_1$ | $\Delta x_2$ | $\Delta x_3$ | $\Delta x_4$ | $\Delta x_5$ | $S_1$ | $S_2$ | $S_3$ | $S_4$ | $S_5$ |
| 1 | 1 | 1 | 1 | 1 | 98 | 101.91 | 151.21 | 10.67 | 9.39 | 9.39 | 8.17 | 8.39 | 1.717 | 1.227 | 0.837 | 2.451[a] | 1.84 |
| 1 | 1 | 1 | 2 | 1 | 86 | 89.63 | 139.13 | 9.38 | 8.64 | 8.64 | 4.30 | 7.68 | 1.247 | 0.904 | 0.617 | 0.520 | 1.336[a] |
| 1 | 1 | 1 | 2 | 2 | 80 | 84.17 | 121.01 | 8.59 | 7.51 | 7.51 | 4.00 | 4.40 | 1.147[a] | 0.805 | 0.549 | 0.347 | 0.205 |
| 2 | 1 | 1 | 2 | 2 | 77 | 74.62 | 106.92 | 5.00 | 6.64 | 6.64 | 3.85 | 3.89 | 0.145 | 0.733[a] | 0.500 | 0.334 | 0.189 |
| 2 | 2 | 1 | 2 | 2 | 71 | 65.06 | 90.80 | 4.25 | 3.72 | 5.64 | 3.55 | 3.30 | 0.124 | 0.068 | 0.426[a] | 0.298 | 0.161 |
| 2 | 2 | 2 | 2 | 2 | 62 | 58.24 | 74.70 | 3.49 | 3.06 | 3.06 | 3.10 | 2.72 | 0.102 | 0.056 | 0.026 | 0.253[a] | 0.133 |
| 2 | 2 | 2 | 3 | 2 | 42 | 45.03 | 56.38 | 2.64 | 2.31 | 2.31 | 1.50 | 2.05 | 0.068 | 0.037 | 0.017 | 0.043 | 0.087[a] |
| 2 | 2 | 2 | 3 | 3 | 32 | 39.15 | 28.89 | 1.35 | 1.18 | 1.18 | 1.06 | 0.71 | 0.040[a] | 0.022 | 0.010 | 0.031 | 0.011 |
| 3 | 2 | 2 | 3 | 3 | 27 | 28.88 | 7.52 | 0.24 | 0.31 | 0.31 | 0.20 | 0.23[b] | | | | | |

[a] This is the stage to which a redundant component is to be added.

[b] Since $L_{+1} = \phi$, stop the procedure.

Table 11    Results of Example 3-4 by Nakagawa and Nakashima's Approach

| Number of components in each stage | | | | | | | | | Stage sensitivity | | |
| $x_1$ | $x_2$ | $x_3$ | $b_1^c$ | $b_2^c$ | $b_3^c$ | $\Delta x_1$ | $\Delta x_2$ | $\Delta x_3$ | $S_1$ | $S_2$ | $S_3$ |
|---|---|---|---|---|---|---|---|---|---|---|---|
| 1 | 1 | 1 | 31.27 | 48.08 | 212.02 | 10.71 | 10.72 | 11.33 | 1.71 | 2.01 | 2.36[a] |
| 1 | 1 | 2 | 29.27 | 41.26 | 199.94 | 8.07 | 9.35 | 9.22 | 1.70[a] | 1.42 | 1.31 |
| 2 | 1 | 2 | 29.86 | 41.11 | 199.94 | 7.21 | 8.17 | 9.01 | 0.27 | 0.88[a] | 0.31 |
| 2 | 2 | 2 | 23.86 | 33.91 | 183.83 | 5.10 | 4.73 | 5.03 | 0.09 | 0.21 | 0.42[a] |
| 2 | 2 | 3 | 21.87 | 26.58 | 159.42 | 3.15 | 3.90 | 4.10 | 0.21[a] | 0.11 | 0.14 |
| 3 | 2 | 3 | 16.13 | 26.41 | 159.42 | 2.33 | 3.10 | 2.97 | 0.04 | 0.09[a] | 0.03 |
| 3 | 3 | 3 | 9.13 | 22.01 | 134.99 | 1.70 | 1.90 | 2.01 | 0.02 | 0.03 | 0.07[a] |
| 3 | 3 | 4 | 9.13 | 14.37 | 96.11 | 0.43 | 1.21 | 1.71 | 0.002 | 0.007 | 0.01[a] |
| 3 | 3 | 5 | 7.12 | 4.68 | 28.2 | 0.31 | 0.27 | 0.27[b] | | | |

[a] This is the stage to which a redundant component is to be added.

[b] Since $L_{+1} = \phi$, stop the procedure.

Table 12    Results of Example 3-2 by the extended Nakagawa and Nakashima's Approach

| Number of components in each subsystem | | | | | Available resource | Stage sensitivity | | | | | | | | | |
|---|---|---|---|---|---|---|---|---|---|---|---|---|---|---|---|
| $x_1$ | $x_2$ | $x_3$ | $x_4$ | $x_5$ | $b^c$ | $\Delta x_1$ | $\Delta x_2$ | $\Delta x_3$ | $\Delta x_4$ | $\Delta x_5$ | $S_1$ | $S_2$ | $S_3$ | $S_4$ | $S_5$ |
| 1 | 1 | 1 | 1 | 1 | 9 | 4.5 | 3.0 | 4.5 | 3.0 | 9.0 | 0.199 | 0.161 | 0.207[a] | 0.096 | 0.032 |
| 1 | 1 | 2 | 1 | 1 | 7 | 3.5 | 2.33 | 3.5 | 2.33 | 7.0 | 0.046 | 0.058 | 0.040 | 0.080[a] | 0.022 |
| 1 | 1 | 2 | 2 | 1 | 4 | 2.0 | 1.33 | 2.0 | 1.33 | 4.0 | 0.035[a] | 0.007 | 0.024 | 0.009 | 0.001 |
| 2 | 1 | 2 | 2 | 1 | 2 | 1.0 | 0.67 | 1.0 | 0.67 | 2.0 | 0.006[a] | | 0.003 | | 0.001 |
| 3 | 1 | 2 | 2 | 1 | $0^b$ | | | | | | | | | | |

[a] This is the stage to which a redundant component is to be added.

[b] Since no resource is available, stop the procedure.

# REFERENCES


1. Aggarwal, K. K., "Redundancy optimization in general systems," *IEEE Transactions on Reliability,* Vol. R-25, No. 5, pp. 330-332 (1976).

2. Aggarwal, K. K., J. S. Gupta, and K. B. Misra, "A new heuristic criterion for solving a redundancy optimization," *IEEE Transactions on Reliability,* Vol. R-24, No. 1, pp. 86-87 (1975).

3. Kuo, W., C. L. Hwang, and F. A. Tillman, "A note on heuristic methods in optimal system reliability," *IEEE Trans. on Reliability,* Vol. R-27, No. 5, pp. 320-324 (1978).

4. Misra, K. B., "A simple approach for constrained redundancy optimization problems," *IEEE Transactions on Reliability,* Vol. R-21, No. 1, pp. 30-34 (1972).

5. Nakagawa, Y., and K. Nakashima, "A heuristic method for determining optimal reliability allocation," *IEEE Transactions on Reliability,* Vol. R-26, No. 3, pp. 156-161 (1977).

6. Sharma, J., and K. V. Venkateswaran, "A direct method for maximizing the system reliability," *IEEE Transactions on Reliability,* Vol. R-20, No. 4, pp. 256-259 (1971).

7. Tillman, F. A., and J. M. Liittschwager, "Integer programming formulation of constrained reliability problems," *Management Science,* Vol. 13, No. 11, pp. 887-899 (1967).

8. Tillman, F. A., "Integer programming solutions to constrained reliability optimization problems," *Transactions of Twentieth Annual Technical Conference,* American Society for Quality Control, Paper No. 66-174, pp. 676-693 (1966).

9. Tillman, F. A., C. L. Hwang, L. T. Fan, and S. A. Balbale, "System reliability subject to multiple nonlinear constraints," *IEEE Transactions on Reliability,* Vol. R-17, No. 3, pp. 153-157 (1968).

10. Ushakov, I. A., "A heuristic method of optimization of the redundancy of multifunction systems," IZV. AN SSSR, Tekhnicheskaya Kibernetika (Engineering Cybernetics) Vol. 10, No. 4, pp. 612-613 (1972).

## 4.1   INTRODUCTION

Dynamic programming is a technique that reduces an N-variable decision
problem to N single-variable problems.  The problem is then structured
as an N-stage sequential decision problem where each single-variable
stage problem is solved sequentially.  Dynamic programming is based
on the "principle of optimality" [1-3] and it yields an exact solution
to the problem. (See Appendix A1).

Various papers have shown how dynamic programming is used to
solve a variety of problems.  Problems solved in these papers are
classified according to the following examples.

*Example 4-1.*  The problem is to locate the redundancy at each stage
of a series system so that the net system profit will be maximized
[17].

Consider an N stage mixed system shown in Fig. 1 having $(x_j - 1)$
parallel redundancies at each stage; the system reliability is

$$R_s = \prod_{j=1}^{N} \left[ 1 - (1 - R_j)^{x_j} \right] \tag{1}$$

Let us assume that P is the profit obtained when the system
operates successfully.  The system reliability, $R_s$, is the fraction
of the trials that are successful and hence the expected profit for
the system is $PR_s$.  Suppose that the costs $C_j$ of the redundant components of the $j^{th}$ stage include the construction costs (suitably

distributed over the life of the process) and the operating costs.
The total cost of the system with redundancies is then

$$\sum_{j=1}^{N} C_j x_j$$

The net profit $N_p$ for the entire system is the profit less the total
cost, that is

$$N_p = PR_s - \sum_{j=1}^{N} C_j x_j \tag{2}$$

The optimal parallel redundancy configuration is the one $x_j$, $j = 1$,
..., N which maximizes the net system profit.

In this example, no constraints are imposed on the problem.

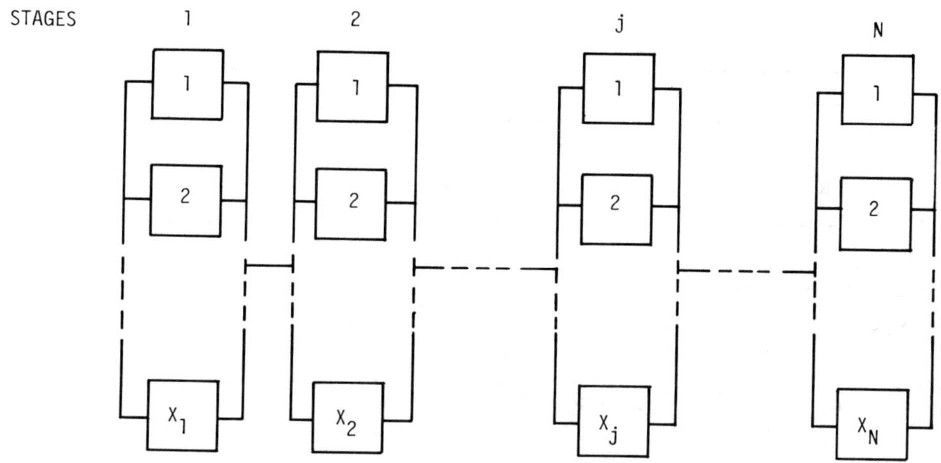


Figure 1    A mixed system with N-stages in series where components
are in parallel at each stage.

Consider a three-stage process.  The profit associated with the final product is P = 10 unit.  The costs, $C_j$, and the reliability $R_j$, of each of the components, are:

|           | $R_j$ | $C_j$ |
|-----------|-------|-------|
| process 3 | 0.333 | 0.20  |
| process 2 | 0.500 | 1.0   |
| process 1 | 0.750 | 1.0   |

*Example 4-2.*  In this problem we assume that a system reliability requirement $R_{s,min}$, is given, and we are to determine a least-cost allocation of an N-stage series system that yields $R_s \geq R_{s,min}$.  The example used is one from Kettelle [10].  Consider the following four-stage system with a system reliability requirement of $R_{s,min} = .99$ and total cost less than $b_1 = 61$:

| Stage   | 4   | 3   | 2    | 1    |
|---------|-----|-----|------|------|
| $C_j$   | 1.2 | 2.3 | 3.4  | 4.5  |
| $R_j$   | 0.8 | 0.7 | 0.75 | 0.85 |

The problem is

Maximize

$$R_s = \prod_{j=1}^{N} \left[ 1 - (1 - R_j)^{x_j} \right]$$

subject to

$$g_1 = \sum_{j=1}^{N} C_j x_j \leq b_1$$

and

$$R_s \geq R_{s,min}$$

*Example 4-3.* This problem is one originally presented by Tillman and Liittschwager [19], where the second constraint in the original problem is excluded. The five-stage problem is stated as

Maximize

$$R_s = \prod_{j=1}^{N} [1 - (1 - R_j)^{x_j}]$$

subject to

$$g_1 = \sum_{j=1}^{N} P_j (x_j)^2 \leq P$$

$$g_2 = \sum_{j=1}^{N} w_j x_j \exp (x_j/4) \leq W$$

where $x_j \geq 1$, $j = 1, 2, \ldots, N$ are integers.

The constants associated with this problem are:

| j | $R_j$ | $P_j$ | P | $w_j$ | W |
|---|-------|-------|-----|-------|-----|
| 1 | 0.80 | 1 |     | 7 |     |
| 2 | 0.85 | 2 |     | 8 |     |
| 3 | 0.90 | 3 | 110 | 8 | 200 |
| 4 | 0.65 | 4 |     | 6 |     |
| 5 | 0.75 | 2 |     | 9 |     |

*Example 4-4.* In this problem we will consider several design alternatives which are available for the $j^{th}$ stage represented by $a_j$ each with a specified inherent component reliability. Let $R'_j(x_j, a_j)$ denote the known reliability function of the $j^{th}$ stage when $x_j$ identical components of design alternative $a_j$ are used. For a N-stage series system, the problem [7] is

Maximize

$$R_x = \prod_{j=1}^{N} R'_j(x_j, a_j)$$

subject to

$$g_1 = \sum_{j=1}^{N} g_{1j}(x_j, a_j) \leq C$$

$$g_2 = \sum_{j=1}^{N} g_{2j}(x_j, a_j) \leq W$$

where

$$x_j \geq 1, \; a_j \geq 1, \; j = 1, 2, \ldots, N \text{ are all integers.}$$

*Example 4-5.* Consider the five-stage problem with three nonlinear constraints in [19] which is

Maximize

$$R_s = \prod_{j=1}^{5} [1 - (1 - R_j)^{x_j}]$$

subject to

$$g_1 = \sum_{j=1}^{N} p_j(x_j)^2 \leq P$$

$$g_2 = \sum_{j=1}^{N} c_j(x_j + \exp(x_j/4)) \leq C$$

$$g_3 = \sum_{j=1}^{N} w_j x_j \exp(x_j/4) \leq W$$

where $x_j$, $j = 1,2,\ldots,N$ are integers.

The objective function $R_s$ can be approximated by

$$R_s \simeq 1 - [(1 - R_1)^{x_1} + (1 - R_2)^{x_2} + (1 - R_3)^{x_3} + (1 - R_4)^{x_4} + (1 - R_5)^{x_5}]$$

where $(1 - R_1)^{x_1}$, $(1 - R_2)^{x_2}$, $(1 - R_3)^{x_3}$, $(1 - R_4)^{x_4}$, and $(1 - R_5)^{x_5}$

are stage unreliabilities and are represented by $Q_1'$, $Q_2'$, $Q_3'$, $Q_4'$, and $Q_5'$, respectively.

The constants associated with the five-stage problem are:

| $j$ | $R_j$ | $p_j$ | $P$ | $c_j$ | $C$ | $w_j$ | $W$ |
|---|---|---|---|---|---|---|---|
| 1 | 0.80 | 1 |     | 7 |     | 7 |     |
| 2 | 0.85 | 2 |     | 7 |     | 8 |     |
| 3 | 0.90 | 3 | 110 | 5 | 175 | 8 | 200 |
| 4 | 0.65 | 4 |     | 9 |     | 6 |     |
| 5 | 0.75 | 2 |     | 4 |     | 9 |     |

By using dynamic programming, the number of constraints will cause the so-called dimensionality difficulty. Three different approaches have been used for solving the problems. They are classified in Table 1.

Table 1   Classification of approaches

| Methods | Application to Examples | References |
|---|---|---|
| Basic dynamic programming approach | Examples 4-1 and 4-2 | 1-4, 6, 7, 11, 12, 13 14, 15, 17 |
| Dynamic programming using Lagrange Multipliers | Examples 4-3, 4-4, and 4-5 | 4, 7, 8, 14 |
| Dynamic programming using the concept of dominating sequence | Examples 4-2 through 4-5 | 9, 10, 16, 20 |

A basic dynamic programming approach is for problems without constraints or those with a single constraint. Whenever there are constraints in a problem, the computation required for solving the problem increases exponentially. The second method in Table 1 was originally introduced by Bellman and Dreyfus [4] where Lagrange multipliers were used when two or more constraints were considered in a problem. By introducing the Lagrange multipliers; the dimensionality problem as a result of having constraints is reduced.

If three constraints are included in a problem, then two Lagrange multipliers have to be introduced, which requires finding two optimal Lagrange multipliers. Therefore, the third approach is suggested which utilizes the concept of dominating sequence (see Table 1). Kettelle [10] apparently is the first one to introduce the concept of dominating sequence to solve a single linear constraint problem. The approach is applicable to problems with 3-nonlinear-constraints. To use this approach we have to find both the upper bound and the lower bound of the number of components used at each stage to reduce the length of the dominating sequence. The detailed discussion with examples are given in the following sections.


## 4.2   BASIC DYNAMIC PROGRAMMING APPROACH

*Solution of Example 4-1.* The basic dynamic programming is used for solving Example 4-1 which is described in the Introduction section.

For the one-stage process, the optimal design is determined for the single decision variable, $x_1$, by the solution of

$$f_1(v_2) = \max_{x_1} \{Pv_1 - C_1 x_1\} \tag{3}$$

for a spectrum of $v_2$ values, where $v_2 = \prod_{j=N+1}^{2} R_{sj}$ is the probability that all upstream stages work, $v_1 = v_2 R_{s1} = v_2[1 - (1 - R_1)^{x_1}]$, and $R_{s,N+1} = v_{N+1} = 1$. $R_{sj}$ is the reliability of the $j^{th}$ stage with $x_j$ parallel components, and $R_j$ is the reliability of each component.

For the two-stage process, the optimal design is obtained by

$$f_2(v_3) = \max_{x_2} \{f_1(v_2) - C_2 x_2\}$$

and for the j-stage process, the recursive functional equation is

$$f_j(v_{j+1}) = \max_{x_j} \{f_{j-1}(v_j) - C_j x_j\} \tag{4}$$

Now, if the optimal design for the subsystem including the stages, N-1, N-2, ..., and 1 is known, then stage N can be designed optimally solving the maximum problem for the single decision variable $x_N$, i.e.,

$$f_N(v_{N+1}) = \max_{x_N} \{f_{N-1}(v_N) - C_N x_N\} \tag{5}$$

Substituting the constants into the equations, the recursive dynamic programming algorithms are

$$f_1(v_2) = \max_{x_1} \{10v_1 - 1.0x_1\} \tag{6}$$

$$f_2(v_3) = \max_{x_2} \{f_1(v_2) - 1.0x_2\} \tag{7}$$

$$f_3(v_4) = \max_{x_3} \{f_2(v_3) - 0.20x_3\} \tag{8}$$

where

$$v_1 = v_2 \{1 - (1 - R_1)^{x_1}\} \tag{9}$$

$$v_2 = v_3 \{1 - (1 - R_2)^{x_2}\} \tag{10}$$

$$v_3 = v_4 \{1 - (1 - R_3)^{x_3}\} \tag{11}$$

$$v_4 = 1.0$$

The first maximum problem (Stage 1) is solved for the optimal $x_1$ for a spectrum of $v_2$ values.  Since $v_2$ is the probability that all upstream stages work, $v_2$ takes a value between 0 and 1.  Equations (6) and (9) are employed in a systematic search for $x_1$ which maximizes $\{10v_1 - 1.0x_1\}$ for an assigned $v_2$ value.  Any one-dimensional search technique can be used; however, since $x_1$ usually takes a small integer value, a simple exhaustive search is carried out and the results are presented in Table 1a. The optimal returns, $f_1(v_2)$, and the optimal parallel components, $x_1$, for $v_2$ = 1.0, 0.9, ..., 0.1 are presented in Table 2 and Fig. 2.  Usually only Table 2 is presented as a dynamic programming table and the detailed calculations presented in Table 1 are omitted.

Similarly equations (7) and (10) are employed in the systematic search for $x_2$ which maximizes $\{f_1(v_2) - C_2 x_2\}$ for each value of $v_3$. The results are presented in Table 1b, and the optimal results in Table 2 and Fig. 2. In the process of calculation, the value of $f_1(v_2)$ is obtained by interpolation.  For example, $v_2$ is given as 0.88 from equation (10) for $v_3$ = 1.0 and $M_2$ = 3.  The value of $f_1(v_2)$ for $v_2$ = 0.88, which is used in equation (7), is determined by interpolation of $f_1(0.9)$ and $f_1(0.8)$ obtained in the stage 1 optimization.  Equations (8) and (11) are used to search for $x_3$ which maximizes $\{f_2(v_3) - C_3 x_3\}$ for $v_4$ = 1.0, since $v_{N+1}$ is always 1.  The results are presented in Table 1c, Table 2 and Fig. 2.

For the three-stage process, the optimal system profit is $f_3$ = 1.32 units with the corresponding optimal values of $x_3$ = 7 and $v_3$ = 0.94.  Entering stage 2 at $v_3$ = 0.94 gives $x_2$ = 3 and $v_2$ = 0.82, and entering stage 1 and $v_2$ = 0.82 gives $x_1$ = 2 and $v_1$ = 0.77.  Thus, the optimal parallel design consists of seven parallel components for stage 3, three parallel components for stage 2, and two parallel

Table 1a   Results of stage 1

| $v_2$ | $x_1$ | $v_1$ | $Pv_1-C_1x_1$ | $f_1(v_2)$ |
|---|---|---|---|---|
| 1.0 | 0 | 0.00 | 0.00 | |
| 1.0 | 1 | 0.75 | 6.50 | |
| 1.0 | 2 | 0.94 | 7.38 | * |
| 1.0 | 3 | 0.98 | 6.84 | |
| 1.0 | 4 | 1.00 | 5.96 | |
| 0.9 | 0 | 0.00 | 0.00 | |
| 0.9 | 1 | 0.68 | 5.75 | |
| 0.9 | 2 | 0.84 | 6.44 | * |
| 0.9 | 3 | 0.89 | 5.86 | |
| 0.9 | 4 | 0.90 | 4.96 | |
| 0.8 | 0 | 0.00 | 0.00 | |
| 0.8 | 1 | 0.60 | 5.00 | |
| 0.8 | 2 | 0.75 | 5.50 | * |
| 0.8 | 3 | 0.79 | 4.88 | |
| 0.8 | 4 | 0.80 | 3.97 | |
| 0.7 | 0 | 0.00 | 0.00 | |
| 0.7 | 1 | 0.53 | 4.25 | |
| 0.7 | 2 | 0.66 | 4.56 | * |
| 0.7 | 3 | 0.69 | 3.89 | |
| 0.7 | 4 | 0.70 | 2.97 | |
| 0.6 | 0 | 0.00 | 0.00 | |
| 0.6 | 1 | 0.45 | 3.50 | |
| 0.6 | 2 | 0.56 | 3.63 | * |
| 0.6 | 3 | 0.59 | 2.91 | |
| 0.6 | 4 | 0.60 | 1.98 | |
| 0.5 | 0 | 0.00 | 0.00 | |
| 0.5 | 1 | 0.38 | 2.75 | * |
| 0.5 | 2 | 0.47 | 2.69 | |
| 0.5 | 3 | 0.49 | 1.92 | |
| 0.4 | 0 | 0.00 | 0.00 | |
| 0.4 | 1 | 0.30 | 2.00 | * |
| 0.4 | 2 | 0.38 | 1.75 | |
| 0.4 | 3 | 0.39 | 0.94 | |
| 0.3 | 0 | 0.00 | 0.00 | |
| 0.3 | 1 | 0.23 | 1.25 | * |
| 0.3 | 2 | 0.28 | 0.81 | |
| 0.3 | 3 | 0.30 | -0.05 | |
| 0.2 | 0 | 0.00 | 0.00 | |
| 0.2 | 1 | 0.15 | 0.50 | * |
| 0.2 | 2 | 0.19 | -0.13 | |
| 0.2 | 3 | 0.20 | -1.03 | |
| 0.1 | 0 | 0.00 | 0.00 | * |
| 0.1 | 1 | 0.07 | -0.25 | |
| 0.1 | 2 | 0.09 | -1.06 | |

Table 1b    Results of stage 2 (and stage 1)

| $v_3$ | $x_2$ | $v_2$ | $f_1(v_2)$ | $C_2 x_2$ | $f_1(v_2) - C_2 x_2$ | $f_2(v_3)$ |
|---|---|---|---|---|---|---|
| 1.0 | 0 | 0.00 | 0.00 | 0.00 | 0.00 | |
| 1.0 | 1 | 0.50 | 2.75 | 1.00 | 1.75 | |
| 1.0 | 2 | 0.75 | 5.03 | 2.00 | 3.03 | |
| 1.0 | 3 | 0.88 | 6.20 | 3.00 | 3.20 | * |
| 1.0 | 4 | 0.94 | 6.79 | 4.00 | 2.79 | |
| 1.0 | 5 | 0.97 | 7.08 | 5.00 | 2.08 | |
| 0.9 | 0 | 0.00 | 0.00 | 0.00 | 0.00 | |
| 0.9 | 1 | 0.45 | 2.38 | 1.00 | 1.38 | |
| 0.9 | 2 | 0.68 | 4.33 | 2.00 | 2.33 | |
| 0.9 | 3 | 0.79 | 5.38 | 3.00 | 2.38 | * |
| 0.9 | 4 | 0.84 | 5.91 | 4.00 | 1.91 | |
| 0.9 | 5 | 0.87 | 6.17 | 5.00 | 1.17 | |
| 0.8 | 0 | 0.00 | 0.00 | 0.00 | 0.00 | |
| 0.8 | 1 | 0.40 | 2.00 | 1.00 | 1.00 | |
| 0.8 | 2 | 0.60 | 3.63 | 2.00 | 1.63 | * |
| 0.8 | 3 | 0.70 | 4.56 | 3.00 | 1.56 | |
| 0.8 | 4 | 0.75 | 5.03 | 4.00 | 1.03 | |
| 0.7 | 0 | 0.00 | 0.00 | 0.00 | 0.00 | |
| 0.7 | 1 | 0.35 | 1.63 | 1.00 | 0.63 | |
| 0.7 | 2 | 0.53 | 2.97 | 2.00 | 0.97 | * |
| 0.7 | 3 | 0.61 | 3.74 | 3.00 | 0.74 | |
| 0.7 | 4 | 0.66 | 4.15 | 4.00 | 0.15 | |
| 0.6 | 0 | 0.00 | 0.00 | 0.00 | 0.00 | |
| 0.6 | 1 | 0.30 | 1.25 | 1.00 | 0.25 | |
| 0.6 | 2 | 0.45 | 2.38 | 3.00 | 0.38 | * |
| 0.6 | 3 | 0.53 | 2.97 | 3.00 | -0.03 | |
| 0.6 | 4 | 0.56 | 3.30 | 4.00 | -0.70 | |
| 0.5 | 0 | 0.00 | 0.00 | 0.00 | 0.00 | * |
| 0.5 | 1 | 0.25 | 0.88 | 1.00 | -0.13 | |
| 0.5 | 2 | 0.38 | 1.81 | 2.00 | -0.19 | |
| 0.4 | 0 | 0.00 | 0.00 | 0.00 | 0.00 | * |
| 0.4 | 1 | 0.20 | 0.50 | 1.00 | -0.50 | |
| 0.4 | 2 | 0.30 | 1.25 | 2.00 | -0.75 | |
| 0.3 | 0 | 0.00 | 0.00 | 0.00 | 0.00 | * |
| 0.3 | 1 | 0.15 | 0.25 | 1.00 | -0.75 | |
| 0.3 | 2 | 0.23 | 0.69 | 2.00 | -1.31 | |
| 0.2 | 0 | 0.00 | 0.00 | 0.00 | 0.00 | * |
| 0.2 | 1 | 0.10 | 0.00 | 1.00 | -1.00 | |
| 0.2 | 2 | 0.15 | 0.25 | 2.00 | -1.75 | |
| 0.1 | 0 | 0.00 | 0.00 | 0.00 | 0.00 | * |
| 0.1 | 1 | 0.05 | 0.00 | 1.00 | -1.00 | |
| 0.1 | 2 | 0.07 | 0.00 | 2,00 | -2.00 | |

# UCL Science Library

www.ucl.ac.uk/library

Borrowed Items 29/01/2010 19:00

XXXXXXX4511

| Item Title | Due Date |
| --- | --- |
| Fatigue of materials / | 03/02/2010 23:5 |
| Control systems: anal | 02/03/2010 23:5 |
| Machine design / J. E | 08/03/2010 23:5 |
| Automotive handbook | 11/03/2010 23:5 |
| Engineering vibration | 16/03/2010 23:5 |
| * Optimization of syste | 26/03/2010 23:5 |

* Indicates items borrowed today

Thank you

www.ucl.ac.uk/library/borrow.shtml

Tel: +44 (0) 20 7679 7795

Table 1c    Results of stage 3 (and stage 2 and stage 1)

| $v_4$ | $x_3$ | $v_3$ | $f_2(v_3)$ | $C_3x_3$ | $f_2(v_3)-C_3x_3$ | $f_3(v_4)$ |
|---|---|---|---|---|---|---|
| 1.0 | 0 | 0.00 | 0.00 | 0.00 | 0.00 | |
| 1.0 | 1 | 0.33 | 0.00 | 0.20 | -0.20 | |
| 1.0 | 2 | 0.56 | 0.21 | 0.40 | -0.19 | |
| 1.0 | 3 | 0.70 | 0.99 | 0.60 | 0.39 | |
| 1.0 | 4 | 0.80 | 1.64 | 0.80 | 0.84 | |
| 1.0 | 5 | 0.87 | 2.14 | 1.00 | 1.14 | |
| 1.0 | 6 | 0.91 | 2.48 | 1.20 | 1.28 | |
| 1.0 | 7 | 0.94 | 2.72 | 1.40 | 1.32 | * |
| 1.0 | 8 | 0.96 | 2.88 | 1.60 | 1.28 | |
| 1.0 | 9 | 0.97 | 2.99 | 1.80 | 1.19 | |

components for stage 1.  This gives rise to the system reliability
of 0.77 at a profit of 1.32 units.  Without parallel redundancy the
system has $P \prod\limits_{j=3}^{1} R_j - \sum\limits_{j=3}^{1} C_j x_j = 10(1.333 \times 0.50 \times 0.75) - (0.20 + 1.0 +$
1.0) = -0.95 profit or cost as the case may be.

*Solution of Example 4-2.*  The problem with a single constraint, Example 4-2, is considered.

The recursive formula of the problem in the basic dynamic programming algorithm is formulated:

$$f_1(b) = \max_{x_1^\ell \leq x_1 \leq x_1^u} [R_1'(x_1)].$$

$$f_2(b) = \max_{x_2^\ell \leq x_2 \leq x_2^u} [R_2'(x_2) \; f_1(b - g_{12}(x_2))]$$

$$\vdots$$

$$f_N(b) = \max_{x_N^\ell \leq x_N \leq x_N^u} [R_N'(x_N) \; f_{N-1}(b - g_{1N}(x_N))]$$

(12)

Table 2    The Dynamic programming table for Example 4-1

stage 3 (and stage 2 and stage 1)

| $v_4$ | $f_3(v_4)$ | $x_3$ | $v_3$ | $f_2(v_3)$ |
|-------|------------|-------|-------|------------|
| 1.0   | 1.32       | 7     | 0.94  | 2.72       |

stage 2 (and stage 1)

| $v_3$ | $f_2(v_3)$ | $x_2$ | $v_2$ | $f_1(v_2)$ |
|-------|------------|-------|-------|------------|
| 1.0   | 3.20       | 3     | 0.88  | 6.20       |
| 0.9   | 2.38       | 3     | 0.79  | 5.38       |
| 0.8   | 1.63       | 2     | 0.60  | 3.63       |
| 0.7   | 0.97       | 2     | 0.53  | 2.97       |
| 0.6   | 0.38       | 2     | 0.45  | 2.38       |
| 0.5   | 0.         | 0     | 0.    | 0.         |
| 0.4   | 0.         | 0     | 0.    | 0.         |
| 0.3   | 0.         | 0     | 0.    | 0.         |
| 0.2   | 0.         | 0     | 0.    | 0.         |
| 0.1   | 0.         | 0     | 0.    | 0.         |

stage 1

| $v_2$ | $f_1(v_2)$ | $x_1$ |
|-------|------------|-------|
| 1.0   | 7.38       | 2     |
| 0.9   | 6.44       | 2     |
| 0.8   | 5.50       | 2     |
| 0.7   | 4.56       | 2     |
| 0.6   | 3.63       | 2     |
| 0.5   | 2.75       | 1     |
| 0.4   | 2.00       | 1     |
| 0.3   | 1.25       | 1     |
| 0.2   | 0.50       | 1     |
| 0.1   | 0.         | 0     |

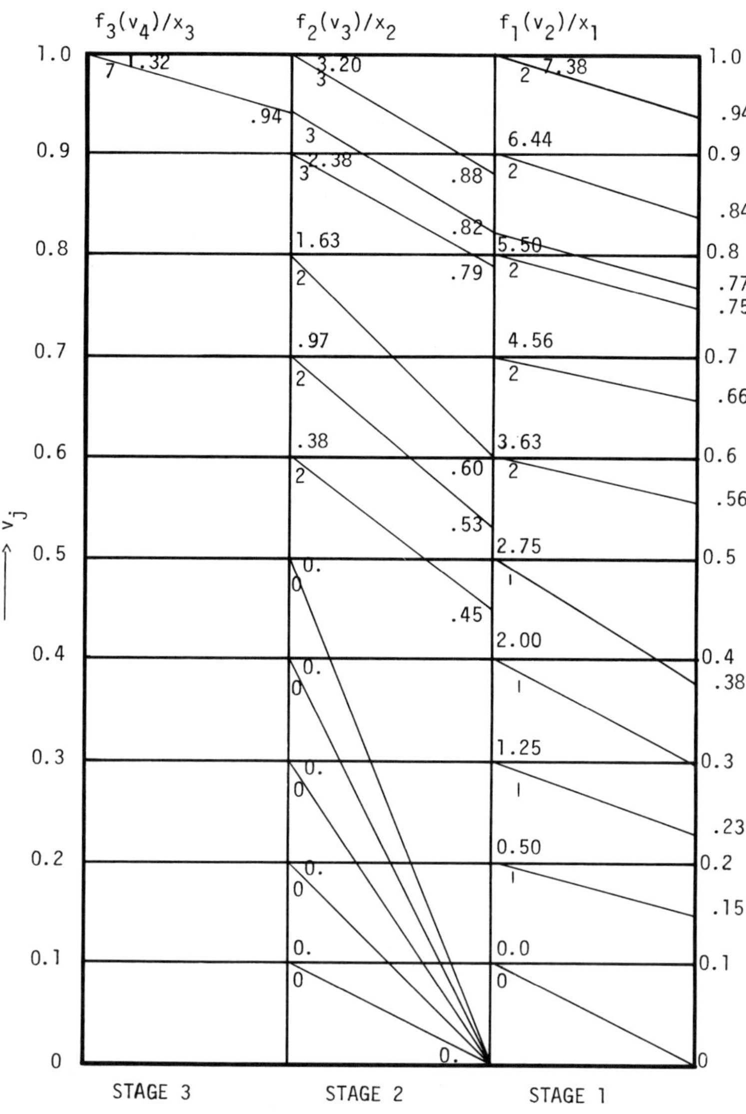


Figure 2    Results for example 4-1.

where

$$R'_j(x_j) = 1 - (1 - R_j)^{x_j}, \qquad j = 1, 2, \ldots, N$$

$x^{\ell}_j$, $j = 1, 2, \ldots, N$ is the minimum integer number used at each stage. Usually $x^{\ell}_j = 1$, $j = 1, 2, \ldots, N$ if no restrictions are on the minimum system reliability. $x^u_j$, $j = 1, 2, \ldots, N$ is the maximum integer number used at each stage such that

$$\sum_{\substack{p=1 \\ p \neq j}}^{N} g_{1p}(x^{\ell}_j) + g_{1j}(x_j) \leq b_1$$

This example has been solved by Kettelle [10] using dynamic programming and the concept of dominating sequences. Here it is solved by the basic dynamic programming approach.

Since the goal for the system reliability is 0.99, the minimum reliability at each stage, is at least 0.99. It is necessary to determine the minimum number of components, $x^{\ell}_j$, used at each stage to attain the stage reliability goal of 0.99. Since the component reliability is 0.85 at stage 1, two components in parallel (one redundancy) give the stage reliability of 0.9775, and three components in parallel (two redundancies) give the stage reliability of 0.9966 which is greater than 0.99. Therefore, three components are the minimum required at stage 1. Similarly, the minimum requirement for stages 2, 3 and 4 are determined to be $(x^{\ell}_2, x^{\ell}_3, x^{\ell}_4) = (4, 4, 3)$, respectively.

Since the maximum of $R_s$ over the feasible region will depend upon the number of stages, N, and the available resource, $b_1$, we denote the maximum $f_N(b_1)$ by $R_N$. That is,

$$f_N(b_1) = \max_{x_N, x_{N-1}, \ldots, x_1} \left[ \prod_{j=N}^{1} R'_j(x_j) \right] \tag{13}$$

where $x_j$, $j = 1, 2, \ldots, N$ are positive integers satisfying the constraint:

$$\sum_{j=1}^{N} g_{1j}(x_j) \leq b_1 \tag{14}$$

For the one-stage process, the optimal design is determined for the single decision variable, $x_1$, by the solution of

$$f_1(b) = \max_{x_1^{\ell} \leq x_1 \leq x_1^{u}} R_1'(x_1) \tag{15}$$

where $x_1^{\ell} = 3$ and the upper bound used in stage 1, $x_1^{u}$, is restricted by the cost constraint. The first maximum problem (stage 1) is solved for the optimal $x_1$ for a spectrum of b values. The spectrum of values for b is determined from the resource consumed, which is, 39.9, for the basic allocation $(x_4, x_3, x_2, x_1) = (3, 4, 4, 3)$, to the total resource available, 61.0. Thus, for each value of b between 39.9 and 61.0, we will find an optimal allocation for $x_1$ when all the upstream stage allocations are fixed by $(x_4^{\ell}, x_3^{\ell}, x_2^{\ell}) = (3, 4, 4)$. The optimal allocation for $x_1$ is shown in Table 3a. All possible b values should be searched exhaustively to find the optimal $x_1$. Since $(x_4, x_3, x_2)$ are fixed, the optimal $x_1$ is 3 for b in the region, $39.90 \leq b \leq 44.40$, which gives $f_1(b) = 0.9768$. For b in the region of $44.40 \leq b \leq 48.90$, the optimal $x_1$ becomes 4 and $f_1(b)$ is 0.9796. Similarly, $x_1 = 5$ for $48.90 \leq b \leq 53.40$, and $f_1(b)$ is 0.9800; $x_1 = 6$ for $53.40 \leq b \leq 57.90$ and $f_1(b)$ is 0.9801; and $x_1 = 7$ for $57.90 \leq b \leq 61.00$ and $f_1(b)$ is 0.9802.

Table 3a   The dynamic programming table of Example 4-2 for Stage 1

| b | $x_4^{\ell}$ | $x_3^{\ell}$ | $x_2^{\ell}$ | $x_1$ | $R_s$ | $f_1(b)$ |
|---|---|---|---|---|---|---|
| 39.90 – 44.39 | 3 | 4 | 4 | 3 | 0.9768 | * |
| 44.40 – 48.89 | 3 | 4 | 4 | 4 | 0.9796 | * |
| 48.90 – 53.39 | 3 | 4 | 4 | 5 | 0.9800 | * |
| 53.40 – 57.89 | 3 | 4 | 4 | 6 | 0.9801 | * |
| 57.90 – 61.00 | 3 | 4 | 4 | 7 | 0.9802 | * |

Table 3b   Computational results of Example 4-2 for Stage 2 (and Stage 1)

| b | $x_4^\ell$ | $x_3^\ell$ | $x_2$ | $x_1$ | $R_s$ | $f_2(b)$ |
|---|---|---|---|---|---|---|
| 40 | 3 | 4 | 4 | 3 | 0.9768 | * |
| 41 | 3 | 4 | 4 | 3 | 0.9768 | * |
| 42 | 3 | 4 | 4 | 3 | 0.9768 | * |
| 43 | 3 | 4 | 4 | 3 | 0.9768 | * |
| 44 | 3 | 4 | 4 | 3 | 0.9768 | |
|  | 3 | 4 | 5 | 3 | 0.9797 | * |
| 45 | 3 | 4 | 5 | 3 | 0.9797 | * |
|  | 3 | 4 | 4 | 4 | 0.9796 | |
| 46 | 3 | 4 | 5 | 3 | 0.9797 | * |
|  | 3 | 4 | 4 | 4 | 0.9796 | |
| 47 | 3 | 4 | 6 | 3 | 0.9804 | * |
|  | 3 | 4 | 4 | 4 | 0.9796 | |
| 48 | 3 | 4 | 6 | 3 | 0.9804 | |
|  | 3 | 4 | 5 | 4 | 0.9825 | * |
| 49 | 3 | 4 | 6 | 3 | 0.9804 | |
|  | 3 | 4 | 5 | 4 | 0.9825 | * |
|  | 3 | 4 | 4 | 5 | 0.9800 | |
| 50 | 3 | 4 | 6 | 3 | 0.9804 | |
|  | 3 | 4 | 5 | 4 | 0.9825 | * |
|  | 3 | 4 | 4 | 5 | 0.9800 | |
| 51 | 3 | 4 | 7 | 3 | 0.9806 | |
|  | 3 | 4 | 5 | 4 | 0.9825 | * |
|  | 3 | 4 | 4 | 5 | 0.9800 | |
| 52 | 3 | 4 | 7 | 3 | 0.9806 | |
|  | 3 | 4 | 6 | 4 | 0.9832 | * |
|  | 3 | 4 | 4 | 5 | 0.9800 | |
| 53 | 3 | 4 | 7 | 3 | 0.9806 | |
|  | 3 | 4 | 6 | 4 | 0.9832 | * |
|  | 3 | 4 | 5 | 5 | 0.9829 | |
| 54 | 3 | 4 | 8 | 3 | 0.9806 | |
|  | 3 | 4 | 6 | 4 | 0.9832 | * |
|  | 3 | 4 | 5 | 5 | 0.9829 | |
|  | 3 | 4 | 4 | 6 | 0.9801 | |
| 55 | 3 | 4 | 8 | 3 | 0.9806 | |
|  | 3 | 4 | 7 | 4 | 0.9834 | * |
|  | 3 | 4 | 5 | 5 | 0.9829 | |
|  | 3 | 4 | 4 | 6 | 0.9801 | |

Table 3b     (continued)

| b | $x_4^\ell$ | $x_3^\ell$ | $x_2$ | $x_1$ | $R_s$ | $f_2(b)$ |
|---|---|---|---|---|---|---|
| 56 | 3 | 4 | 8 | 3 | 0.9806 | |
|    | 3 | 4 | 7 | 4 | 0.9834 | |
|    | 3 | 4 | 6 | 5 | 0.9836 | * |
|    | 3 | 4 | 4 | 6 | 0.9801 | |
| 57 | 3 | 4 | 9 | 3 | 0.9806 | |
|    | 3 | 4 | 7 | 4 | 0.9834 | |
|    | 3 | 4 | 6 | 5 | 0.9836 | * |
|    | 3 | 4 | 5 | 6 | 0.9830 | |
| 58 | 3 | 4 | 9 | 3 | 0.9806 | |
|    | 3 | 4 | 8 | 4 | 0.9835 | |
|    | 3 | 4 | 6 | 5 | 0.9836 | * |
|    | 3 | 4 | 5 | 6 | 0.9830 | |
|    | 3 | 4 | 4 | 7 | 0.9802 | |
| 59 | 3 | 4 | 9 | 3 | 0.9806 | |
|    | 3 | 4 | 8 | 4 | 0.9835 | |
|    | 3 | 4 | 6 | 5 | 0.9836 | * |
|    | 3 | 4 | 5 | 6 | 0.9830 | |
|    | 3 | 4 | 4 | 7 | 0.9802 | |
| 60 | 3 | 4 | 9 | 3 | 0.9806 | |
|    | 3 | 4 | 8 | 4 | 0.9835 | |
|    | 3 | 4 | 7 | 5 | 0.9838 | * |
|    | 3 | 4 | 5 | 6 | 0.9830 | |
|    | 3 | 4 | 4 | 7 | 0.9802 | |
| 61 | 3 | 4 | 10 | 3 | 0.9806 | |
|    | 3 | 4 | 8 | 4 | 0.9835 | |
|    | 3 | 4 | 7 | 5 | 0.9838 | * |
|    | 3 | 4 | 6 | 6 | 0.9837 | |
|    | 3 | 4 | 4 | 7 | 0.9802 | |

For all possible b we have searched the optimal $x_1$'s when stages 4, 3, and 2 are fixed at $x_4^\ell$, $x_3^\ell$, $x_2^\ell$. The next step is to search for the optimal combinations of Stage 2 and Stage 1 when Stage 4 and Stage 3 are fixed at the minimum required components, $x_4^\ell$, and $x_3^\ell$. It is still necessary to consider all possible b between 39.90 and 61.0. For convenience, the maximization will be carried out for b from 40.0 to 61.0 with a discrete spectrum. The difference between every two near searching point is one. Table 3a then is used to construct Table 3b.

Table 3c    Computational results of Example 4-2 for Stage 3 (and Stage 2 and Stage 1)

| b | $x_4^{\ell}$ | $x_3$ | $x_2$ | $x_1$ | $R_s$ | $f_3(b)$ |
|---|---|---|---|---|---|---|
| 40 | 3 | 4 | 4 | 3 | 0.9768 | * |
| 41 | 3 | 4 | 4 | 3 | 0.9768 | * |
| 42 | 3 | 4 | 4 | 3 | 0.9768 | * |
| 43 | 3 | 5 | 4 | 3 | 0.9824 | * |
| 44 | 3 | 5 | 4 | 3 | 0.9824 | * |
|    | 3 | 4 | 5 | 3 | 0.9797 |   |
| 45 | 3 | 6 | 4 | 3 | 0.9841 | * |
|    | 3 | 4 | 5 | 3 | 0.9797 |   |
| 46 | 3 | 6 | 4 | 3 | 0.9841 |   |
|    | 3 | 5 | 5 | 3 | 0.9853 | * |
| 47 | 3 | 7 | 4 | 3 | 0.9846 |   |
|    | 3 | 5 | 5 | 3 | 0.9853 | * |
|    | 3 | 4 | 6 | 3 | 0.9804 |   |
| 48 | 3 | 7 | 4 | 3 | 0.9846 |   |
|    | 3 | 6 | 5 | 3 | 0.9870 | * |
|    | 3 | 4 | 6 | 3 | 0.9804 |   |
|    | 3 | 4 | 5 | 4 | 0.9825 |   |
| 49 | 3 | 7 | 4 | 3 | 0.9846 |   |
|    | 3 | 6 | 5 | 3 | 0.9870 | * |
|    | 3 | 5 | 6 | 3 | 0.9860 |   |
|    | 3 | 4 | 5 | 4 | 0.9825 |   |
| 50 | 3 | 8 | 4 | 3 | 0.9847 |   |
|    | 3 | 6 | 5 | 3 | 0.9870 | * |
|    | 3 | 5 | 6 | 3 | 0.9860 |   |
|    | 3 | 4 | 5 | 4 | 0.9825 |   |
| 51 | 3 | 8 | 4 | 3 | 0.9847 |   |
|    | 3 | 7 | 5 | 3 | 0.9875 |   |
|    | 3 | 5 | 6 | 3 | 0.9860 |   |
|    | 3 | 5 | 5 | 4 | 0.9881 | * |
| 52 | 3 | 9 | 4 | 3 | 0.9848 |   |
|    | 3 | 7 | 5 | 3 | 0.9875 |   |
|    | 3 | 6 | 6 | 3 | 0.9877 |   |
|    | 3 | 5 | 5 | 4 | 0.9881 | * |
|    | 3 | 4 | 6 | 4 | 0.9832 |   |
| 53 | 3 | 9 | 4 | 3 | 0.9848 |   |
|    | 3 | 8 | 5 | 3 | 0.9876 |   |
|    | 3 | 6 | 6 | 3 | 0.9877 |   |
|    | 3 | 6 | 5 | 4 | 0.9898 | * |
|    | 3 | 4 | 6 | 4 | 0.9832 |   |

Table 3c    (continued)

| b | $x_4^{\ell}$ | $x_3$ | $x_2$ | $x_1$ | $R_s$ | $f_3(b)$ |
|---|---|---|---|---|---|---|
| 54 | 3 | 10 | 4 | 3 | 0.9848 | |
|    | 3 | 8 | 5 | 3 | 0.9876 | |
|    | 3 | 7 | 6 | 3 | 0.9882 | |
|    | 3 | 6 | 5 | 4 | 0.9898 | * |
|    | 3 | 5 | 6 | 4 | 0.9888 | |
| 55 | 3 | 10 | 4 | 3 | 0.9848 | |
|    | 3 | 9 | 5 | 3 | 0.9877 | |
|    | 3 | 7 | 6 | 3 | 0.9882 | |
|    | 3 | 7 | 5 | 4 | 0.9903 | * |
|    | 3 | 5 | 6 | 4 | 0.9888 | |
|    | 3 | 4 | 7 | 4 | 0.9834 | |
| 56 | 3 | 11 | 4 | 3 | | |
|    | 3 | 9 | 5 | 3 | 0.9877 | |
|    | 3 | 8 | 6 | 3 | 0.9883 | |
|    | 3 | 7 | 5 | 4 | 0.9903 | |
|    | 3 | 6 | 6 | 4 | 0.9905 | * |
|    | 3 | 4 | 6 | 5 | 0.9836 | |
| 57 | 3 | 11 | 4 | 3 | | |
|    | 3 | 9 | 5 | 3 | 0.9877 | |
|    | 3 | 8 | 6 | 3 | 0.9883 | |
|    | 3 | 8 | 5 | 4 | 0.99046 | |
|    | 3 | 6 | 6 | 4 | 0.99053 | * |
|    | 3 | 4 | 6 | 5 | 0.9836 | |
| 58 | 3 | 12 | 4 | 3 | | |
|    | 3 | 10 | 5 | 3 | 0.9877 | |
|    | 3 | 8 | 6 | 3 | 0.9883 | |
|    | 3 | 8 | 5 | 4 | 0.99046 | |
|    | 3 | 6 | 6 | 4 | 0.99053 | * |
|    | 3 | 5 | 6 | 5 | 0.9893 | |
| 59 | 3 | 12 | 4 | 3 | | |
|    | 3 | 10 | 5 | 3 | 0.9877 | |
|    | 3 | 9 | 6 | 3 | 0.9884 | |
|    | 3 | 8 | 5 | 4 | 0.99046 | |
|    | 3 | 7 | 6 | 4 | 0.9910 | * |
|    | 3 | 5 | 6 | 5 | 0.9893 | |
| 60 | 3 | 13 | 4 | 3 | 0.9849 | |
|    | 3 | 11 | 5 | 3 | 0.9877 | |
|    | 3 | 9 | 6 | 3 | 0.9884 | |
|    | 3 | 9 | 5 | 4 | 0.9905 | |
|    | 3 | 7 | 6 | 4 | 0.9910 | * |
|    | 3 | 5 | 6 | 5 | 0.9893 | |
|    | 3 | 4 | 7 | 5 | 0.9838 | |

Table 3c     (continued)

| b | $x_4^\ell$ | $x_3$ | $x_2$ | $x_1$ | $R_s$ | $f_3(b)$ |
|---|---|---|---|---|---|---|
| 61 | 3 | 13 | 4 | 3 | 0.9849 | |
| | 3 | 11 | 5 | 3 | 0.9877 | |
| | 3 | 10 | 6 | 3 | 0.9884 | |
| | 3 | 9 | 5 | 4 | 0.9905 | |
| | 3 | 8 | 6 | 4 | 0.9912 | * |
| | 3 | 6 | 6 | 5 | 0.9910 | |
| | 3 | 4 | 7 | 5 | 0.9838 | |

Table 3d   Computational results of Example 4-2 for Stage 4
(and Stage 3, Stage 2 and Stage 1)

| b | $x_4$ | $x_3$ | $x_2$ | $x_1$ | $R_s$ | $f_4(b)$ |
|---|---|---|---|---|---|---|
| 61 | 20 | 4 | 4 | 3 | 0.9848 | |
| | 18 | 5 | 4 | 3 | 0.9904 | |
| | 16 | 6 | 4 | 3 | 0.9921 | |
| | 15 | 5 | 5 | 3 | 0.9933 | |
| | 14 | 7 | 4 | 3 | 0.9926 | |
| | 13 | 6 | 5 | 3 | 0.9951 | |
| | 11 | 5 | 5 | 4 | 0.99609 | |
| | 10 | 6 | 5 | 4 | 0.99779 | |
| | 8 | 7 | 5 | 4 | 0.99830 | |
| | 7 | 6 | 6 | 4 | 0.99851 | |
| | 5 | 7 | 6 | 4 | 0.99871 | |
| | 3 | 8 | 6 | 4 | 0.99119 | |

In Table 3b, $(x_4, x_3) = (3, 4)$ are always fixed. An optimal $(x_2, x_1)$ is searched for the maximum system reliability for the corresponding value of b given. For example, if b = 40, from Table 3a, $x_1$ could be 3; with $(x_4, x_3, x_1) = (3, 4, 3)$, the optimal $x_2$ is 4; then the system reliability for $(x_4, x_3, x_2, x_1) = (3, 4, 4, 3)$ is 0.9768. Similarly, when b = 41, 42, and 43, the optimal allocation for $(x_2, x_1)$ are (4, 3). When b is increased to 44, from Table 3a, $x_1$ is still 3, but $x_2$ can be 4 or 5, and $x_2 = 5$ gives the greater system reliability of 0.9797. Therefore, $f_2(44) = 0.9797$.

Table 4    The dynamic programming table of Example 4-2

Stage 4 (and Stage 3, Stage 2 and Stage 1)

| b | $x_4$ | $x_3$ | $x_2$ | $x_1$ | $f_4(b)$ |
|---|---|---|---|---|---|
| 61 | 5 | 7 | 6 | 4 | 0.99871 |

Stage 3 (and stage 2 and stage 1)

| b | $x_4^\ell$ | $x_3$ | $x_2$ | $x_1$ | $f_3(b)$ |
|---|---|---|---|---|---|
| 40 | 3 | 4 | 4 | 3 | 0.9768 |
| 41 | 3 | 4 | 4 | 3 | 0.9768 |
| 42 | 3 | 4 | 4 | 3 | 0.9768 |
| 43 | 3 | 5 | 4 | 3 | 0.9824 |
| 44 | 3 | 5 | 4 | 3 | 0.9824 |
| 45 | 3 | 6 | 4 | 3 | 0.9841 |
| 46 | 3 | 5 | 5 | 3 | 0.9853 |
| 47 | 3 | 5 | 5 | 3 | 0.9855 |
| 48 | 3 | 6 | 5 | 3 | 0.9870 |
| 49 | 3 | 6 | 5 | 3 | 0.9870 |
| 50 | 3 | 6 | 5 | 3 | 0.9870 |
| 51 | 3 | 5 | 5 | 4 | 0.9881 |
| 52 | 3 | 5 | 5 | 4 | 0.9881 |
| 53 | 3 | 6 | 5 | 4 | 0.9898 |
| 54 | 3 | 6 | 5 | 4 | 0.9898 |
| 55 | 3 | 7 | 5 | 4 | 0.9903 |
| 56 | 3 | 6 | 6 | 4 | 0.9905 |
| 57 | 3 | 6 | 6 | 4 | 0.9905 |
| 58 | 3 | 6 | 6 | 4 | 0.9905 |
| 59 | 3 | 7 | 6 | 4 | 0.9910 |
| 60 | 3 | 7 | 6 | 4 | 0.9910 |
| 61 | 3 | 8 | 6 | 4 | 0.9912 |

Table 4   Table 4 (continued)

Stage 2 (and stage 1)

| b | $x_4^\ell$ | $x_3^\ell$ | $x_2$ | $x_1$ | $f_2(b)$ |
|---|---|---|---|---|---|
| 40 | 3 | 4 | 4 | 3 | 0.9768 |
| 41 | 3 | 4 | 4 | 3 | 0.9768 |
| 42 | 3 | 4 | 4 | 3 | 0.9768 |
| 43 | 3 | 4 | 4 | 3 | 0.9768 |
| 44 | 3 | 4 | 5 | 3 | 0.9797 |
| 45 | 3 | 4 | 5 | 3 | 0.9797 |
| 46 | 3 | 4 | 5 | 3 | 0.9797 |
| 47 | 3 | 4 | 6 | 3 | 0.9804 |
| 48 | 3 | 4 | 5 | 4 | 0.9825 |
| 49 | 3 | 4 | 5 | 4 | 0.9825 |
| 50 | 3 | 4 | 5 | 4 | 0.9825 |
| 51 | 3 | 4 | 5 | 4 | 0.9825 |
| 52 | 3 | 4 | 6 | 4 | 0.9832 |
| 53 | 3 | 4 | 6 | 4 | 0.9831 |
| 54 | 3 | 4 | 6 | 4 | 0.9831 |
| 55 | 3 | 4 | 7 | 4 | 0.9834 |
| 56 | 3 | 4 | 6 | 5 | 0.9836 |
| 57 | 3 | 4 | 6 | 5 | 0.9836 |
| 58 | 3 | 4 | 6 | 5 | 0.9836 |
| 59 | 3 | 4 | 6 | 5 | 0.9836 |
| 60 | 3 | 4 | 7 | 5 | 0.9838 |
| 61 | 3 | 4 | 7 | 5 | 0.9838 |

Stage   1

| b | $x_4^\ell$ | $x_3^\ell$ | $x_2^\ell$ | $x_1$ | $f_1(b)$ |
|---|---|---|---|---|---|
| 39.90 − 44.39 | 3 | 4 | 4 | 3 | 0.9768 |
| 44.40 − 48.89 | 3 | 4 | 4 | 4 | 0.9796 |
| 48.90 − 53.39 | 3 | 4 | 4 | 5 | 0.9800 |
| 53.40 − 57.89 | 3 | 4 | 4 | 6 | 0.9807 |
| 57.90 − 61.00 | 3 | 4 | 4 | 7 | 0.9802 |

When b is increased to 45, then from Table 3a, $x_1$ can be either 3 or 4. When $x_1$ = 3, we search for the optimal $x_2$ to be 5 and $R_s$ = 0.9797; when $x_1$ = 4, we search for the optimal $x_2$ to be 4 and $R_s$ = 0.9796, the optimal allocation for b = 45 is $(x_4, x_3, x_2, x_1)$ = (3, 4, 5, 3). The computational results presented in Table 3b are similarly carried out. Consider another example, for b = 54, from Table 3a, $x_1$ can be 3, 4, 5, or 6 as $(x_4, x_3)$ are fixed as (3, 4). For $x_1$ = 3, the maximum system reliability is obtained when $x_2$ = 8. Similarly for $x_1$ = 4, the maximum system reliability is given at $x_2$ = 6; for $x_1$ = 5 at $x_2$ = 5; for $x_1$ = 6 at $x_2$ = 4. The optimum result for b = 54, $f_2(54)$, is the maximum system reliability among $(x_2, x_1)$ = (8, 3), (6, 4), (5, 5), and (4, 6), which is $R_s$ = 0.9832 and $(x_2^*, x_1^*)$ = (6, 4). Usually the computational results for stage 2 (and stage 1) presented in Table 3b are not presented and only the dynamic programming table of Table 4 is presented.

Similarly, we can construct Table 3c for all possible b values and for fixed $x_4$ = 3. For each b value, a systematic search procedure is carried out by looking back at the optimal allocation of $(x_2, x_1)$s shown in Table 4 for stage 2. For example, when b = 52 is of interest, from Table 4, the various optimal allocations for $(x_2, x_1)$, for b $\leq$ 52 are:

|   | Optimum | |
|---|---|---|
| b | $(x_2$ | $x_1)$ |
| 40 | 4 | 3 |
| 41 | 4 | 3 |
| 42 | 4 | 3 |
| 43 | 4 | 3 |
| 44 | 5 | 3 |
| 45 | 5 | 3 |
| 46 | 5 | 3 |
| 47 | 6 | 3 |
| 48 | 5 | 4 |
| 49 | 5 | 4 |
| 50 | 5 | 4 |
| 51 | 5 | 4 |
| 52 | 6 | 4 |

Therefore, the optimal allocation for $(x_2, x_1)$ can only be one of the following: (4, 3), (5, 3), (6, 3), (5, 4), (6, 4). Since $x_4$ is fixed, for $(x_2, x_1) = (4, 3)$ we find the optimal is $x_3 = 9$ with the system reliability, $R_s = 0.9848$; for $(x_2, x_1) = (5, 3)$, the optimal is $x_3 = 7$ and $R_s = 0.9875$; for $(x_2, x_1) = (6, 3)$ the optimal is $x_3 = 6$, and $R_s = 0.9877$; for $(x_2, x_1) = (5, 4)$, the optimal is $x_3 = 5$ and $R_s = 0.9881$; and for $(x_2, x_1) = (6, 4)$ the optimal is $x_3 = 4$ and $R_s = 0.9832$. Among these system reliabilities, 0.9881 is the largest one, hence the allocation of $(x_4, x_3, x_2, x_1) = (3, 5, 5, 4)$ is the optimal one for b = 52. The optimum results for stage 3 (and stage 2 and stage 1) are presented in Table 4.

Finally we can construct Table 3d for b = 61, which is the total available resource. For b = 61, from Table 4 for stage 3, all the optimal allocations for $(x_3, x_2, x_1)$ for $b \leq 61$ are: (4, 4, 3), (5, 4, 3), (6, 4, 3), (5, 5, 5), (6, 5, 3), (5, 5, 4), (6, 5, 4), (7, 5, 4), (6, 6, 4), (7, 6, 4), and (8, 6, 4). For each allocation, the optimum $x_4$ (the maximum allowable $x_4$ to give the maximum system reliability) is calculated. Among all these system reliabilities, the optimal system reliability for this problem as shown in Table 3d, is $(x_4, x_3, x_2, x_1) = (5, 7, 6, 4)$ which gives the largest system reliability, $R_s = 0.99871$. The dynamic programming table for this problem is given in Table 4.


## 4.3 DYNAMIC PROGRAMMING APPROACH USING LAGRANGE MULTIPLIERS

Formulation of the problem

If multiple constraint functions are imposed on the objective function, then Lagrange multipliers may be introduced to eliminate some constraints and hence reduce the dimensions of the problem.

In Section 4.2, we have formulated the single constraint problem which was solved by the basic dynamic programming approach. Now, if the second constraint,

$$\sum_{j=1}^{N} g_{2j}(x_j) \leq b_2 \tag{16}$$

is also imposed on the problem, we have to consider the sequence of functions defined by the relation

$$f_1(b_1, b_2) = \max_{1 \leq x_1 \leq x_1^u} R_1'(x_1) \tag{17}$$

$$f_2(b_1, b_2) = \max_{1 \leq x_2 \leq x_2^u} [R_2'(x_2) * f_1(b_1 - g_{12}(x_2), b_2 - g_{22}(x_2))]$$

$$\vdots$$

$$f_N(b_1, b_2) = \max_{1 \leq x_N \leq x_N^u} [R_N'(x_N) * f_{N-1}(b_1 - g_{1N}(x_N), b_2 - g_{2N}(x_N))]$$

where $x_j^u$, $j = 1, 2, \ldots, N$ is the minimum integer between $(x_j^u)^1$ and $(x_j^u)^2$; and $(x_j^u)^1$ is the maximum integer satisfying

$$\sum_{\substack{p=1 \\ p \neq j}}^{N} g_{1p}(1) + g_{1j}(x_j) \leq b_1$$

and $(x_j^u)^2$ is the maximum integer satisfying

$$\sum_{\substack{p=1 \\ p \neq j}}^{N} g_{2p}(1) + g_{2j}(x_j) \leq b_2$$

The recursive formula for a two-constraints problem is basically the same approach as that used for a one-constraint problem.  Although the formula is simple and straightforward, it involves sequences of functions of two variables which will require a large memory capacity and are quite time-consuming. Therefore, it is not very desirable from a computational standpoint.

An alternative method to solve this problem is by introducing a Lagrange multiplier, $\lambda$, as a penalty term. The problem now is stated as

Maximize

$$\prod_{j=1}^{N} R_j'(x_j) \exp\left[-\lambda \sum_{j=1}^{N} g_{2j}(x_j)\right] \tag{18}$$

subject to

$$\sum_{j=1}^{N} g_{1j}(x_j) \le b_1$$

The Lagrange multiplier, $\lambda$, is to be chosen so that the constraint in eq. (16) is as close to being equal as possible. Now the problem becomes one of a sequence of functions of one variable each which has the following recursive formula:

$$f_1(b) = \max_{x_1^{\ell} \le x_1 \le x_1^{u}} [R_1'(x_1) \exp(-\lambda g_{21}(x_1))]$$

$$f_2(b) = \max_{x_2^{\ell} \le x_2 \le x_2^{u}} [R_2'(x_2) f_1(b_1 - g_{12}(x_2)) \exp(-\lambda g_{22}(x_2))]$$

.

.

.

$$f_N(b) = \max_{x_N^{\ell} \le x_N \le x_N^{u}} [R_N'(x_N) f_{N-1}(b_1 - g_{1N}(x_N)) \exp(-\lambda g_{2N}(x_N))]$$

where

$$R_j'(x_j) = 1 - (1 - R_j)^{x_j}, \qquad j = 1,2,\ldots,N$$

$x_j^{\ell}$, $j = 1,2,\ldots,N$ is the minimum integer number used at each stage

and $x_j^{u}$, $j = 1,2,\ldots,N$ is the maximum integer number used at each stage such that

$$\sum_{\substack{p=1 \\ p \neq j}}^{N} g_{1p}(x_j^{\ell}) + g_{1j}(x_j) \leq b_1$$

$\lambda$ is to be chosen so that $\sum_{j=1}^{N} g_{2j}(x_j)$ is as close to $b_2$ as possible.
For a fixed value of $\lambda$, the maximum system reliability is obtained;
that is,

$$R_s = f_N(b_1) \exp\left[-\lambda \sum_{j=1}^{N} g_{2j}(x_j)\right]$$

A one-dimensional search for $\lambda$ should be carried out to find the optimal solution for $R_s$.


*Solution of example 4-3*  To solve this example, we first find the
lower bound of components to be used at each stage.

By an enumeration method as shown in Table 5, redundancies are
allocated stage by stage until one of the constraints is exceeded.
The basic system configuration for calculating the lower bound is assumed to be the one before the point of exceeding one of the constraints,
i.e., $(3, 3, 3, 2, 2)$, for this numerical example. The system reliability corresponding to this configuration, $R_s(\bar{x})$, is 0.8125.  This
is not, however, an optimal solution.  The optimal system reliability
should be equal to or greater than this value.  Therefore, we assume
that the lower bound of the stage reliability is 0.8125 and we calculate the corresponding lower bound of stage components.  That is, for
$j = 1$,

$$1 - (1 - 0.80)^{x_1} \geq 0.8125$$

which gives $x_1 \geq 2.31$, say $x_1^{\ell} = 2$.  Similarly we obtain $x_2^{\ell} = 2$, $x_3^{\ell} = 1$, $x_4^{\ell} = 3$, and $x_5^{\ell} = 2$.

Table 5    The allocation of elements stage by stage until
any one constraint is violated.

| STAGES | | | | | RESOURCES USED | |
|---|---|---|---|---|---|---|
| 1 | 2 | 3 | 4 | 5 | $\sum\limits_{j=1}^{5} g_{1j}$ | $\sum\limits_{j=1}^{5} g_{2j}$ |
| 1 | 1 | 1 | 1 | 1 | 12 | 48.79 |
| 2 | 1 | 1 | 1 | 1 | 15 | 62.88 |
| 2 | 2 | 1 | 1 | 1 | 21 | 78.99 |
| 2 | 2 | 2 | 1 | 1 | 30 | 95.10 |
| 2 | 2 | 2 | 2 | 1 | 42 | 107.19 |
| 2 | 2 | 2 | 2 | 2 | 48 | 125.30 |
| 3 | 2 | 2 | 2 | 2 | 53 | 146.67 |
| 3 | 3 | 2 | 2 | 2 | 63 | 171.11 |
| 3 | 3 | 3 | 2 | 2 | 78 | 195.53 |
| 3 | 3 | 3 | 3 | 2 | 98 | 213.86 |

The recursive equations modified by using the Lagrange multiplier
are

$$f_1(b) = \max_{x_1^{\ell} \le x_1 \le x_1^{u}} \{(1 - Q_1^{x_1}) \exp[-\lambda(w_1 x_1 e^{x_1/4})]\} \tag{19}$$

$$f_2(b) = \max_{x_2^{\ell} \le x_2 \le x_2^{u}} (1 - Q_2^{x_2}) \exp[-\lambda(w_2 x_2 e^{x_2/4})] \, f_1(b - p_2 x_2^2) \tag{20}$$

$$f_3(b) = \max_{x_3^{\ell} \le x_3 \le x_3^{u}} (1 - Q_3^{x_3}) \exp[-\lambda(w_3 x_3 e^{x_3/4})] \, f_2(b - p_3 x_3^2) \tag{21}$$

$$f_4(b) = \max_{x_4^{\ell} \le x_4 \le x_4^{u}} (1 - Q_4^{x_4}) \exp[-\lambda(w_4 x_4 e^{x_4/4})] \, f_3(b - p_4 x_4^2) \tag{22}$$

$$f_5(b) = \max_{x_5^{\ell} \le x_5 \le x_5^{u}} (1 - Q_5^{x_5}) \exp[-\lambda(w_5 x_5 e^{x_5/4})] \, f_4(b - p_5 x_5^2) \tag{23}$$

where $Q_j = (1 - R_j)$, $j = 1, 2, \ldots, 5$.

The quantity $\lambda$ is to be determined so that

$$g_2 = \sum_{j=1}^{N} w_j x_j \exp\,(x_j/4) \simeq W$$

To solve this example, $\lambda$ should be assigned, say $\lambda = 0.001$. Since the objective over the feasible region depends upon the number of stages, $N$, the available resource, $b_1$, and the Lagrange multiplier, $\lambda$, we denote by $f_N(b_1)$ the maximization of the objective. That is

$$f_N(b_1) = \max_{x_N, x_{N-1}, \ldots, x_1} \left[\, \prod_{j=N}^{1} R_j'(x_j) \exp\,\left(-\lambda \sum_{j=1}^{N} g_{2j}(x_j)\right)\right]$$

where $x_j$, $j = 1,2,\ldots,N$ are positive integers satisfying the constraint

$$\sum_{j=1}^{N} g_{1j}(x_j) \le b_1$$

For the one-stage process, the optimal design is determined for the single decision variable, $x_1$, by the solution of

$$f_1(b) = \max_{x_1^{\ell} \le x_1 \le x_1^{u}} R_1'(x_1) \exp\,(-\lambda g_{2j}(x_1))$$

where, $x_1^{\ell} = 2$, and the upper bound used in stage 1, $x_1^{\ell}$, are restricted by the constraint. The spectrum of b is determined from the consumed resource, 59.0, for the basic allocation $(x_5^{\ell},\ x_4^{\ell},\ x_3^{\ell},\ x_2^{\ell},\ x_1^{\ell}) = (2, 3, 1, 2, 2)$ to the total available resource of 110.0. When b increases, stage redundancy, $x_1 - 1$, stage reliability, $R_1'(x_1)$, and stage cost, $g_{21}(x_1)$, will increase, but the penalty term, $\exp\,(-\lambda g_{21}(x_1))$, will decrease. Since $f_1(b)$ is a maximization of the product of $R_1'(x_1)$ and $\exp\,(-\lambda g_{21}(x_1))$, $f_1(b)$ is not a monotonically increasing function of b. In other words, the increase in b allows us to add more components in stage 1 but the configuration from this increased redundancy may not give us an optimal return value. When the upstream stages are fixed by $(x_5^\ell, x_4^\ell, x_3^\ell, x_2^\ell) = (2, 3, 1, 2)$, the optimal allocation for $x_1$ is obtained as shown in Table 6a. All possible b values should be searched exhaustively to find the optimal $x_1$. Since $(x_5, x_4, x_3, x_2)$ are fixed, $x_1$ is 2 for b in the region of $59.0 \le b < 64.0$ which gives the functional value of 0.66710. For b in the region of $64.0 \le b < 71.0$, the optimal $x_1$ becomes 3 and $f_1(b)$ is 0.67476. For b in the region of $71.0 \le b < 80.0$, we may allocate 4 components for $x_1$, but this gives the functional value of 0.65860 which is smaller than the functional value of 0.67476 as $x_1 = 3$. Therefore, $x_1 = 3$ is the optimal one for $71.0 \le b < 80.0$. Similarly, for $b \ge 80.0$, we may allocate $x_1 = 3, 4, 5, 6, 7$; however, $f_1(b)$, the optimum is at $x_1 = 3$. Therefore, the optimal is $x_1 = 3$ for $64.0 \le b < 110$.

The results for stage 1 are presented in the dynamic programming table (Table 7, stage 1).

The next step is to search for the optimal combination of values for stage 2 and stage 1 since $(x_5^\ell, x_4^\ell, x_3^\ell) = (2, 3, 1)$ is fixed. It is necessary to consider all possible values of b between 59.0 and 110.0. For convenience, the maximization will be carried out for b with a discrete spectrum. Since the cost of adding one more component to the minimum required components of any stage will consume at least 5 cost units, the difference between two near searching points can be chosen as 5. Table 7 (stage 1) then is used to construct Table 6b.

In Table 6b, $(x_5^\ell, x_4^\ell, x_3^\ell)$ is $(2, 3, 1)$ is fixed. An optimal $(x_2, x_1)$ is searched for the maximum value of the function $f_2(b)$, for the corresponding value of b given. This procedure is similar to the one given in the basic dynamic programming algorithm. For example, if $b = 84$, from Table 7 (stage 1) $x_1$ can be 2 or 3. When $x_1 = 2$, we search for the optimal $x_2$ to be 2, and the function value is 0.66710; when $x_1 = 3$, we search for the optimal $x_2$ to be 2 and the functional value is 0.67476. Since 0.67476 is greater than 0.66710, the optimal allocation for $b = 84$ is $(x_5^\ell, x_4^\ell, x_3^\ell, x_2, x_1) = (2, 3, 1, 2, 3)$. In Table 6b, when $(x_5^\ell, x_4^\ell, x_3^\ell) = (2, 3, 1)$ is fixed, only two possible allocations exist for $x_2$ and $x_1$, namely, $(x_2, x_1) = (2, 2)$ or $(2, 3)$, which are presented in the dynamic programming table, Table 7 (stage 2). Similarly, we can construct Table 6c for stage 3 (and stage 2 and stage 1) for all possible b values and for fixed $(x_5^\ell, x_4^\ell) = (2, 3)$.

Table 6a   Calculated Results of Example 4-3 for Stage 1 as $\lambda = 0.0010$.

| b | $x_5^{\ell}$ | $x_4^{\ell}$ | $x_3^{\ell}$ | $x_2^{\ell}$ | $x_1$ | functional value | $f_1^{\ell}(b)$ |
|---|---|---|---|---|---|---|---|
| 59–63.9 | 2 | 3 | 1 | 2 | 2 | 0.66710 | * |
| 64–70.9 | 2 | 3 | 1 | 2 | 3 | 0.67476 | * |
| 71–79.9 | 2 | 3 | 1 | 2 | 3 | 0.67476 | * |
|  | 2 | 3 | 1 | 2 | 4 | 0.65860 |  |
| 80–90.9 | 2 | 3 | 1 | 2 | 3 | 0.67476 | * |
|  | 2 | 3 | 1 | 2 | 4 | 0.65860 |  |
|  | 2 | 3 | 1 | 2 | 5 | 0.62915 |  |
| 91–103.9 | 2 | 3 | 1 | 2 | 3 | 0.67476 | * |
|  | 2 | 3 | 1 | 2 | 4 | 0.65860 |  |
|  | 2 | 3 | 1 | 2 | 5 | 0.62915 |  |
|  | 2 | 3 | 1 | 2 | 6 | 0.60322 |  |
| 104–110 | 2 | 3 | 1 | 2 | 3 | 0.67476 | * |
|  | 2 | 3 | 1 | 2 | 4 | 0.65860 |  |
|  | 2 | 3 | 1 | 2 | 5 | 0.62915 |  |
|  | 2 | 3 | 1 | 2 | 6 | 0.60322 |  |
|  | 2 | 3 | 1 | 2 | 7 | 0.58209 |  |

For each b value, a systematic search procedure is carried out by using the previously determined optimal allocation of $(x_2, x_1)$ shown in Table 7 (stage 2). We can also construct Table 6d for stage 4 (and stage 3, stage 2, and stage 1) for all possible b values and for fixed $x_5^{\ell} = 2$.

Finally we can construct Table 6e for b = 110 which is the total available resource. For b = 110, from Table 7 (stage 4), all the possible optimal allocations of $(x_4, x_3, x_2, x_1)$ for $b \leq 110$ are: (3, 1, 2, 2), (3, 1, 2, 3), (3, 2, 2, 2), and (3, 2, 2, 3). For each allocation, the optimum $x_5$ (the maximum allowable $x_5$ to give the maximum functional value) is calculated, and from all these possible values we choose the largest one as the optimal one. As shown in Table 6e, $(x_5, x_4, x_3, x_2, x_1) = (3, 3, 2, 2, 3)$ gives $f_5(b = 110) = 0.74610$. The dynamic programming table for $\lambda = 0.001$ is given in Table 7.

Table 6b  Calculated Results of Example 4-3 for Stage 2 (and Stage 1) as $\lambda = 0.0010$.

| b | $x_5^{\ell}$ | $x_4^{\ell}$ | $x_3^{\ell}$ | $x_2$ | $x_1$ | functional value | $f_2(b)$ |
|---|---|---|---|---|---|---|---|
| 59 | 2 | 3 | 1 | 2 | 2 | 0.66710 | * |
| 64 | 2 | 3 | 1 | 2 | 2 | 0.66710 | |
|    | 2 | 3 | 1 | 2 | 3 | 0.67476 | * |
| 69 | 2 | 3 | 1 | 2 | 2 | 0.66710 | |
|    | 2 | 3 | 1 | 2 | 3 | 0.67476 | * |
| 74 | 2 | 3 | 1 | 2 | 2 | 0.66710 | |
|    | 2 | 3 | 1 | 2 | 3 | 0.67476 | * |
| 79 | 2 | 3 | 1 | 2 | 2 | 0.66710 | |
|    | 2 | 3 | 1 | 2 | 3 | 0.67476 | * |
| 84 | 2 | 3 | 1 | 2 | 2 | 0.66710 | |
|    | 2 | 3 | 1 | 2 | 3 | 0.67476 | * |
| 89 | 2 | 3 | 1 | 2 | 2 | 0.66710 | |
|    | 2 | 3 | 1 | 2 | 3 | 0.67476 | * |
| 94 | 2 | 3 | 1 | 2 | 2 | 0.66710 | |
|    | 2 | 3 | 1 | 2 | 3 | 0.67476 | * |
| 99–110 | 2 | 3 | 1 | 2 | 2 | 0.66710 | |
|        | 2 | 3 | 1 | 2 | 3 | 0.67476 | * |

Table 8 shows that when $\lambda = 0.001$, we have $f_5(b = 110) = 0.74610$.

The total consumed $g_2 = \sum_{j=1}^{N} g_{2j}(x_j) = 192.5$.  The system reliability, $R_s$, is 0.9045, which is given by

$$f_5(b)/\exp\left(-\lambda \sum_{j=1}^{N} g_{2j}(x_j)\right).$$

In searching for the proper value of the Lagrange multiplier, $\lambda$, which spends a cost close to 200 (but always less than 200 since $g_2 \leq W$, where $W = 200$), several values of $\lambda$ have been tried.  For each value of $\lambda$, the procedure presented above is carried out and an optimum configuration is obtained.  The results are summarized in Table 8.

Table 6c    Calculated Results of Example 4-3 for Stage 3 (and Stage 2 and Stage 1) as $\lambda = 0.0010$.

| b | $x_5^\ell$ | $x_4^\ell$ | $x_3$ | $x_2$ | $x_1$ | functional value | $f_3(b)$ |
|---|---|---|---|---|---|---|---|
| 59 | 2 | 3 | 1 | 2 | 2 | 0.66710 | * |
| 64 | 2 | 3 | 1 | 2 | 2 | 0.66710 | |
| | 2 | 3 | 1 | 2 | 3 | 0.67476 | * |
| 69 | 2 | 3 | 2 | 2 | 2 | 0.72208 | * |
| | 2 | 3 | 1 | 2 | 3 | 0.67476 | |
| 74 | 2 | 3 | 2 | 2 | 2 | 0.72208 | |
| | 2 | 3 | 2 | 2 | 3 | 0.73037 | * |
| 79 | 2 | 3 | 2 | 2 | 2 | 0.72208 | |
| | 2 | 3 | 2 | 2 | 3 | 0.73037 | * |
| 84 | 2 | 3 | 2 | 2 | 2 | 0.72208 | |
| | 2 | 3 | 2 | 2 | 3 | 0.73037 | * |
| 89 | 2 | 3 | 2 | 2 | 2 | 0.72208 | |
| | 2 | 3 | 2 | 2 | 3 | 0.73037 | * |
| 94 | 2 | 3 | 2 | 2 | 2 | 0.72208 | |
| | 2 | 3 | 2 | 2 | 3 | 0.73037 | * |
| 99–110 | 2 | 3 | 2 | 2 | 2 | 0.72208 | |
| | 2 | 3 | 2 | 2 | 3 | 0.73037 | * |

For $\lambda = 0.0001$, the optimal allocation is $(x_5, x_4, x_3, x_2, x_1) = (4, 3, 2, 3, 3)$, which gives the system reliability, $R_s = 0.9331$, and consumes $g_1 = 107$, and $g_2 = 257.6$, i.e., the second constraint is violated. For $\lambda = 0.01$, the optimal allocation is $(x_5, x_4, x_3, x_2, x_1) = (2, 3, 1, 2, 2)$, which gives the system reliability, $R_s = 0.7578$, and consumes $g_1 = 59$, and $g_2 = 127.5$. Now 127.5 is smaller than 200 and the solution is a feasible one. However, we can increase the stage redundancies, and consume more resource to increase the system reliability. The one-dimensional search for $\lambda$ can be applied between 0.0001 and 0.01. Table 8 gives the optimal solution: $0.0008 \le \lambda \le 0.0015$, $(x_5, x_4, x_3, x_2, x_1) = (3, 3, 2, 2, 3)$, $g_1 = 83$, $g_2 = 192.5$, and $R_s = 0.9045$.

Table 6d    Calculated Results of Example 4-3 for Stage 4 (and Stage 3, Stage 2 and Stage 1) as $\lambda = 0.0010$.

| b | $x_5^{\ell}$ | $x_4$ | $x_3$ | $x_2$ | $x_1$ | functional value | $f_4(b)$ |
|---|---|---|---|---|---|---|---|
| 59 | 2 | 3 | 1 | 2 | 2 | 0.66710 | * |
| 64 | 2 | 3 | 1 | 2 | 2 | 0.66710 | |
|    | 2 | 3 | 1 | 2 | 3 | 0.67476 | * |
| 69 | 2 | 3 | 1 | 2 | 2 | 0.66710 | |
|    | 2 | 3 | 1 | 2 | 3 | 0.67476 | |
|    | 2 | 3 | 2 | 2 | 2 | 0.72208 | * |
| 74 | 2 | 3 | 1 | 2 | 2 | 0.66710 | |
|    | 2 | 3 | 1 | 2 | 3 | 0.67476 | |
|    | 2 | 3 | 2 | 2 | 2 | 0.72208 | |
|    | 2 | 3 | 2 | 2 | 3 | 0.73037 | * |
| 79-110 | 2 | 3 | 1 | 2 | 2 | 0.66710 | |
|    | 2 | 3 | 1 | 2 | 3 | 0.67476 | |
|    | 2 | 3 | 2 | 2 | 2 | 0.72208 | |
|    | 2 | 3 | 2 | 2 | 3 | 0.73037 | * |

Table 6e    Calculated Results of Example 4-3 for Stage 5 (and Stage 4, Stage 3, Stage 2 and Stage 1) as $\lambda = 0.001$.

| b | $x_5$ | $x_4$ | $x_3$ | $x_2$ | $x_1$ | function value | $f_5(b)$ |
|---|---|---|---|---|---|---|---|
| 110 | 3 | 3 | 1 | 2 | 2 | 0.68147 | |
|     | 3 | 3 | 1 | 2 | 3 | 0.68929 | |
|     | 3 | 3 | 2 | 2 | 2 | 0.73764 | |
|     | 3 | 3 | 2 | 2 | 3 | 0.74610 | * |

Table 7    The Dynamic Programming Table for Example 4-3 as $\lambda = 0.0010$.

Stage 5 (and Stage 4, Stage 3, Stage 2 and Stage 1)

| b | $x_5$ | $x_4$ | $x_3$ | $x_2$ | $x_1$ | $f_5(b)$ |
|---|---|---|---|---|---|---|
| 110 | 3 | 3 | 2 | 2 | 3 | 0.74610 |

Stage 4 (and Stage 3, Stage 2 and Stage 1)

| b | $x_5^{\ell}$ | $x_4$ | $x_3$ | $x_2$ | $x_1$ | $f_4(b)$ |
|---|---|---|---|---|---|---|
| 59 | 2 | 3 | 1 | 2 | 2 | 0.66710 |
| 64 | 2 | 3 | 1 | 2 | 3 | 0.67476 |
| 69 | 2 | 3 | 2 | 2 | 2 | 0.72208 |
| 74-110 | 2 | 3 | 2 | 2 | 3 | 0.73037 |

Stage 3 (and Stage 2 and Stage 1)

| b | $x_5^{\ell}$ | $x_4^{\ell}$ | $x_3$ | $x_2$ | $x_1$ | $f_3(b)$ |
|---|---|---|---|---|---|---|
| 59 | 2 | 3 | 1 | 2 | 2 | 0.66710 |
| 64 | 2 | 3 | 1 | 2 | 3 | 0.67476 |
| 69 | 2 | 3 | 2 | 2 | 2 | 0.72208 |
| 74-110 | 2 | 3 | 2 | 2 | 3 | 0.73037 |

Stage 2 (and Stage 1)

| b | $x_5^{\ell}$ | $x_4^{\ell}$ | $x_3^{\ell}$ | $x_2$ | $x_1$ | $f_2(b)$ |
|---|---|---|---|---|---|---|
| 59 | 2 | 3 | 1 | 2 | 2 | 0.66710 |
| 64-110 | 2 | 3 | 1 | 2 | 3 | 0.67476 |

Table 7    (continued)

| b | $x_5^{\ell}$ | $x_4^{\ell}$ | $x_3^{\ell}$ | $x_2^{\ell}$ | $x_1$ | $f_1(b)$ |
|---|---|---|---|---|---|---|
| 56-63.9 | 2 | 3 | 1 | 2 | 2 | 0.66710 |
| 64-110 | 2 | 3 | 1 | 2 | 3 | 0.67476 |

Table 8    Optimum system reliabilities for various values of Lagrange multiplier.

| Lagrange Multiplier | Optimum System Configuration | | | | | Optimum System Reliability | | |
|---|---|---|---|---|---|---|---|---|
| $\lambda$ | $x_5$ | $x_4$ | $x_3$ | $x_2$ | $x_1$ | $R_s$ | $g_1$ | $g_2$ |
| 0.0001 | 4 | 3 | 2 | 3 | 3 | 0.9331 | 107 | 257.6 |
| 0.0002 | 4 | 3 | 2 | 3 | 3 | 0.9331 | 107 | 257.6 |
| 0.0004 | 4 | 3 | 2 | 3 | 3 | 0.9331 | 107 | 257.6 |
| 0.0006 | 3 | 3 | 2 | 3 | 3 | 0.9222 | 93 | 216.9 |
| 0.0008 | 3 | 3 | 2 | 2 | 3 | 0.9045 | 83 | 192.5 |
| 0.0010 | 3 | 3 | 2 | 2 | 3 | 0.9045 | 83 | 192.5 |
| 0.0015 | 3 | 3 | 2 | 2 | 3 | 0.9045 | 83 | 192.5 |
| 0.0016 | 3 | 3 | 2 | 2 | 2 | 0.8753 | 78 | 171.1 |
| 0.0040 | 2 | 3 | 2 | 2 | 2 | 0.8336 | 68 | 143.6 |
| 0.0060 | 2 | 3 | 1 | 2 | 2 | 0.7578 | 59 | 127.5 |
| 0.0080 | 2 | 3 | 1 | 2 | 2 | 0.7578 | 59 | 127.5 |
| 0.0100 | 2 | 3 | 1 | 2 | 2 | 0.7578 | 59 | 127.5 |

## 4.4  DYNAMIC PROGRAMMING APPROACH USING THE CONCEPT OF DOMINATING SEQUENCES

Formulation of the problem

The number of computations required for maximizing the system reliability

$$R_s = \prod_{j=1}^{N} [1 - (1 - R_j)^{x_j}]$$

subject to

$$g_i = \sum_{j=1}^{N} g_{ij}(x_j) \le b_i, \qquad i = 1,2,\ldots,r \qquad (24)$$

can be reduced by defining a condition of dominance for alternative system configurations.

A system configuration $\bar{x}'$ is said to dominate another system configuration $\bar{x}$, if

$$R_s(\bar{x}') \ge R_s(\bar{x})$$

and the inequality sign ($<$) holds in at least one of the following conditions

$$\sum_{j=1}^{N} g_{ij}(\bar{x}'_j) \le \sum_{j=1}^{N} g_{ij}(\bar{x}_j), \qquad i = 1,2,\ldots,r$$

This implies that the dominating system configuration has better system reliability while using less cost (resources).  A sequence S of redundancy allocations, all satisfying the constraints in (24) and none dominated by the others, is said to form a dominating sequence.

In the dynamic programming formulation, combinations of two stages are searched for a dominating sequence of configurations which is then combined with a third stage to yield another dominating sequence. A sequence ends whenever a constraint is violated.  The final dominant

configuration yielding the optimal system configuration is the last
entry in the dominating sequence process generated by the combination
of the dominating sequence from stage 1, stage 2, ..., stage N-1, and
stage N.

To reduce the length of the dominating sequence, heuristic techniques are used to determine the upper and lower bounds of $x_j$, $j = 1,$
$2,...,N$.

i)   Upper bound of $x_j$, $x_j^u$:

Each stage should have at least one component.  If the upper
bound of the $j^{th}$ stage, $x_j^u$, is to be determined, we let

$$x_k = 1, \qquad k = 1,2,..,N, \qquad k \neq j$$

$x_j^u$ is the smallest integer number in the set $\{c_1, c_2, ..., c_r\}$, where
$c_\ell = \max \{x_j | x_j$ is integer, and $g_{\ell k} (1,...,1,x_j,1...,1) \leq b_\ell\}$ for
$\ell = 1,2,...,r$.

ii) Lower bound of $x_j$, $x_j^\ell$:

Redundancies are allocated stage by stage until a constraint is
met.  If the reliability of the configuration $\bar{x}$, which is the last
step of the allocation process, is $R_s(\bar{x})$ and does not violate any
constraint, then N equations of the form $R_s(\bar{x}) \leq 1 - (1 - R_j)^{x_j}$ are
solved for $x_j$, where $x_j^\ell$ is the set of minimum integer numbers satisfying the above equations for $j = 1,2,...,N$.  $x_j^\ell$ is the lower bound
for the components used at stage j.


*Solution of example 4-5.*  To use the concept of dominating sequences
to solve this example, we first must find the upper and lower bounds
for the components used at each stage.

i)   Upper bound, $x_j^u$:

To find the upper bounds of components for the $j^{th}$ stage, all
the other stages are assumed to have one component.  The upper bound
for the first stage, $x_1^u$, will be the largest integer number satisfying
the following three constraints:

$$g_1 = 1 * (x_1^u)^2 + 2 * (1)^2 + 3 * (1)^2 + 4 * (1)^2 + 2 * (1)^2 \leq 110$$

$$g_2 = 7 * (x_1^u + \exp (x_1^u/4)) + 7 * (1 + \exp (1/4))$$

$$+ 5 * (1 + \exp (1/4))$$

$$+ 9 * (1 + \exp (1/4)) + 4 * (1 + \exp (1/4)) \leq 175$$

$$g_3 = 7 * x_1^u \exp (x_1^u/4) + 8 * 1 * \exp (1/4) + 8 * 1 * \exp (1/4)$$

$$+ 6 * 1 * \exp (1/4) + 9 * 1 * \exp (1/4) \leq 200$$

By inserting the integer number, $x_1^u = 1,2,\ldots$ into $g_1$, $g_2$, $g_3$, when $x_1^u = 6$, we have $g_1 = 47$, $g_2 = 137.34$, and $g_3 = 227.36$; that is, $g_3(x_1^u = 6)$ is greater than 200. When $x_1^u = 5$, however, $g_1 = 36$, $g_2 = 91.44$, and $g_3 = 161.59$. None of the constraints are violated. Therefore, $x_1^u$ is 5. By similar procedures, the upper bounds for the components for the other stages are all found to be 5.

ii)  Lower bound, $x_j^{\ell}$:

By an enumeration method as shown in Table 9, redundancies are allocated stage by stage until one of the constraints is exceeded. The basic system configuration for calculating the lower bound, is assumed to be the one just before a constraint is exceeded, i.e., (3, 3, 3, 2, 2), for this numerical example.

The system reliability corresponding to this configuration of (3, 3, 3, 2, 2), $R_s(\bar{x})$, is 0.8124. This is, however, not an optimal solution. The optimal system reliabilty should be equal to or greater than this value. Therefore, we assume that the lower bound of stage reliability is 0.8124, and calculate the corresponding lower bound for the stage components. That is, for j=1:

$$1 - (1 - 0.80)^{x_1} \geq 0.8124$$

gives $x_1 \geq 2.31$, say $x_1^{\ell} = 2$. Similarly we obtain $x_2^{\ell} = 2$, $x_3^{\ell} = 1$, $x_4^{\ell} = 3$, and $x_5^{\ell} = 2$.

Table 9   The allocation of elements stage by stage until any
one constraint is violated.

| STAGES | | | | | RESOURCES USED | | |
| --- | --- | --- | --- | --- | --- | --- | --- |
| 1 | 2 | 3 | 4 | 5 | $\sum_{j=1}^{5} g_{1j}$ | $\sum_{j=1}^{5} g_{2j}$ | $\sum_{j=1}^{5} g_{3j}$ |
| 1 | 1 | 1 | 1 | 1 | 12 | 73.09 | 48.79 |
| 2 | 1 | 1 | 1 | 1 | 15 | 82.64 | 62.88 |
| 2 | 2 | 1 | 1 | 1 | 21 | 92.19 | 78.99 |
| 2 | 2 | 2 | 1 | 1 | 30 | 99.01 | 95.10 |
| 2 | 2 | 2 | 2 | 1 | 42 | 111.29 | 107.18 |
| 2 | 2 | 2 | 2 | 2 | 48 | 116.75 | 125.30 |
| 3 | 2 | 2 | 2 | 2 | 53 | 127.03 | 146.67 |
| 3 | 3 | 2 | 2 | 2 | 63 | 138.31 | 171.11 |
| 3 | 3 | 3 | 2 | 2 | 78 | 144.65 | 195.53 |
| 3 | 3 | 3 | 3 | 2 | 98 | 157.87 | 213.86 |

The optimum configuration for the components at each stage will
then lie between the lower and upper bounds of that stage.

To solve this example, the first step in the computational procedure is to set up a matrix for the combination of stage 1 and stage
2 (see Table 10a.)   In Table 10a the number of components, the stage
unreliability, and $g_1$, $g_2$, and $g_3$ for stages 1 and 2 are presented as
the rows above the matrix and the column to the left of the matrix,
respectively.   The starting number of components used for each stage
is the lower bound for the stage, and the ending point, the upper
bound.   It is easier to consider unreliabilities than reliabilities,
although it involves an approximation.

Each entry of the matrix in Table 10a is a vector, which shows
the system unreliabilities, and $g_1$, $g_2$, and $g_3$ which are results of
the combination of stages 1 and 2.   The system unreliability is approximated by the addition of the unreliabilities for stages 1 and 2, if
both $R_1$ and $R_2$ are near unity, namely, $R_1$ and $R_2 \geq 0.5$,

$$Q' = 1 - (1 - (1 - R_1)^{x_1})(1 - (1 - R_2)^{x_2})$$

$$\simeq (1 - R_1)^{x_1} + (1 - R_2)^{x_2}$$

where $(1 - R_1)^{x_1}$ and $(1 - R_2)^{x_2}$ are the unreliabilities of stage 1 and stage 2, respectively.

The dominating sequence for the system by combining stage 1 with stage 2 is obtained by eliminating entries of the matrix which are dominated by others. These eliminating procedures are:

(1)  When any cost of the entries in the matrix exceeds the constrained available resource, the entry is eliminated. For example, the entries of $(x_1, x_2) = (5, 4)$, $(4, 5)$, and $(5, 5)$ are eliminated, because all $g_3$'s of the entries exceed 200.

(2) The dominating sequence will then be determined as follows:

a.  Consider the entry having the highest reliability (i.e., the lowest unreliability), which is always one term of the dominating sequence no matter what the costs are. In Table 10a, this entry is $(x_1, x_2) = (4, 4)$, which has the highest reliability, $1 - 0.0021 = 0.9979$. Now compare the costs of all the other entries with costs of this entry. Eliminate all entries which have a lower reliability and a higher cost. In Table 10a, the highest entry is $(x_1, x_2) = (4, 4)$, which has a reliability of 0.9979, $g_1 = 48$, $g_2 = 94.04$, and $g_3 = 163.08$ Comparing with $(x_1, x_2) = (4, 4)$, the entry $(x_1, x_2) = (3, 5)$, which has a reliability of $1 - 0.0081 = 0.9919$, $g_1 = 59$, $g_2 = 95.17$, and $g_3 = 183.66$, is eliminated, because the latter is less reliable and requires higher cost for $g_1$, $g_2$, and $g_3$. That is, entry (4, 4) dominates entry (3, 5).

b.  Choose the next higher reliability (lower unreliability), i.e., entry (5, 3). Compare the costs of all other entries which have a lower reliability than entry (5, 3). However, no entry is dominated by (5, 3).

c.  Next, entries (3, 4) and (2, 5) are eliminated by comparing with entry (4, 3); entry (2, 4) and (5, 2) are eliminated by comparing with entry (3, 3); and entry (2, 3) is eliminated by comparing with

Table 10a    Computational results of Example 4-5 for Stage 1 and Stage 2.

**Stage 1**

| Number of components used | $x_1^\ell = 2$ | 3 | 4 | $x_1^u = 5$ |
|---|---|---|---|---|
| Stage unreliability | 0.04 | 0.008 | 0.0016 | 0.0003 |
| $g_1$ used | 4 | 9 | 16 | 25 |
| $g_2$ used | 25.54 | 35.81 | 47.02 | 59.36 |
| $g_3$ used | 23.08 | 44.45 | 76.10 | 121.81 |

**Stage 2**

| $x_2^\ell = 2$ | | (1) | (2) | (3) | |
|---|---|---|---|---|---|
| 0.0225 | 0.0625 | 0.0305 | 0.0241 | 0.0228 |
| 8 | 12 | 17 | 24 | 33 |
| 25.54 | 51.08 | 61.35 | 72.56 | 84.90 |
| 26.38 | 49.46 | 70.83 | 102.48 | 148.19 |

| 3 | | | (4) | (5) | (6) |
|---|---|---|---|---|---|
| 0.0034 | 0.0434 | 0.0114 | 0.0050 | 0.0037 |
| 18 | 22 | 27 | 34 | 43 |
| 35.81 | 61.35 | 71.62 | 82.83 | 95.17 |
| 50.81 | 73.89 | 95.26 | 126.91 | 172.61 |

| 4 | | | | (7) | |
|---|---|---|---|---|---|
| 0.0005 | 0.0405 | 0.0085 | 0.0021 | 0.008 |
| 32 | 36 | 41 | 48 | 57 |
| 47.02 | 72.56 | 82.83 | 94.04 | 106.38 |
| 86.98 | 110.06 | 131.43 | 163.08 | 208.79 |

| $x_2^u = 5$ | | | | | |
|---|---|---|---|---|---|
| 0.0001 | 0.0401 | 0.0081 | 0.0017 | 0.0004 |
| 50 | 54 | 59 | 66 | 75 |
| 59.36 | 84.90 | 95.17 | 106.38 | 118.72 |
| 139.21 | 162.29 | 183.66 | 215.31 | 261.02 |

Table 10b  Computational Results of Example 4-5 for Stage 1-2 and Stage 3

| | Stage 1-2 | | | | | | |
|---|---|---|---|---|---|---|---|
| Number of components used | 2-2 | 3-2 | 4-2 | 3-3 | 4-3 | 5-3 | 4-4 |
| Stage unreliability | 0.0625 | 0.0305 | 0.0241 | 0.0114 | 0.0050 | 0.0037 | 0.0021 |
| $g_1$ used | 12 | 17 | 24 | 27 | 34 | 43 | 48 |
| $g_2$ used | 51.08 | 61.35 | 72.56 | 71.62 | 82.83 | 95.17 | 94.04 |
| $g_3$ used | 49.46 | 70.83 | 102.48 | 95.26 | 126.91 | 172.62 | 163.08 |
| $x_3^{\ell} = 1$  0.10 | 0.1625 [1] | 0.1305 [2] | 0.1241 | 0.1114 | 0.1050 | 0.1037 | 0.1021 |
|  3 | 15 | 20 | 27 | 30 | 37 | 46 | 51 |
|  11.42 | 62.50 | 72.76 | 83.98 | 83.04 | 94.25 | 106.59 | 105.46 |
|  10.27 | 59.73 | 81.10 | 112.75 | 105.53 | 137.28 | 182.89 | 173.35 |
| $x_3 = 2$  0.01 | 0.0725 [3] | 0.0405 [5] | 0.0341 [6] | 0.0214 [7] | 0.0150 [8] | 0.0137 | 0.0121 [10] |
|  12 | 24 | 29 | 36 | 39 | 46 | 55 | 60 |
|  18.24 | 69.32 | 79.59 | 90.80 | 89.86 | 101.07 | 113.41 | 112.28 |
|  26.38 | 75.84 | 97.21 | 128.86 | 121.64 | 153.29 | 199.00 | 189.46 |
| $x_3 = 3$  0.001 | 0.0635 [4] | 0.0315 | 0.0251 | 0.0124 [9] | 0.0060 [12] | 0.0047 | 0.0031 |
|  27 | 39 | 44 | 51 | 54 | 61 | 70 | 75 |
|  25.58 | 76.66 | 89.93 | 98.14 | 97.20 | 108.41 | 120.75 | 119.02 |
|  50.81 | 100.27 | 121.64 | 153.29 | 146.07 | 177.72 | 223.43 | 213.89 |
| $x_3^{u} = 4$  0.0001 | 0.0626 | 0.0306 | 0.0242 | 0.0115 [11] | 0.0052 | 0.0038 | 0.0022 |
|  48 | 60 | 65 | 72 | 75 | 82 | 91 | 96 |
|  33.58 | 84.66 | 94.93 | 106.14 | 105.20 | 116.41 | 128.75 | 127.61 |
|  86.98 | 136.44 | 157.81 | 189.46 | 182.24 | 213.89 | 259.60 | 250.06 |

Table 10c   Computational results of Example 4-5 for Stage 1-2-3 and Stage 4

**Stage 1-2-3**

| Number of components used | | 2-2-1 | 3-2-1 | 2-2-2 | 2-2-3 | 3-2-2 | 4-2-2 | 3-3-2 | 4-3-2 | 3-3-3 | 4-4-2 | 3-3-4 | 4-3-3 |
|---|---|---|---|---|---|---|---|---|---|---|---|---|---|
| Stage unreliability | | 0.1625 | 0.1305 | 0.0725 | 0.0635 | 0.0405 | 0.0341 | 0.0214 | 0.0150 | 0.0124 | 0.0121 | 0.0115 | 0.0060 |
| $g_1$ used | | 15 | 20 | 24 | 39 | 29 | 36 | 39 | 46 | 54 | 60 | 75 | 61 |
| $g_2$ used | | 62.50 | 72.76 | 69.32 | 76.66 | 79.59 | 90.80 | 89.86 | 101.07 | 97.20 | 112.28 | 105.20 | 108.41 |
| $g_3$ used | | 59.73 | 81.10 | 75.84 | 100.27 | 97.21 | 128.86 | 121.64 | 153.29 | 146.07 | 189.46 | 182.24 | 177.72 |
| $x_4^{\ell} = 3$ | | (1) | (2) | (3) | (4) | (5) | (6) | (7) | (8) | (10) | | | |
| | 0.0429 | 0.2054 | 0.1734 | 0.1154 | 0.1064 | 0.0834 | 0.0770 | 0.0643 | 0.0579 | 0.0553 | 0.0550 | 0.0544 | 0.0489 |
| | 36 | 51 | 56 | 60 | 75 | 65 | 72 | 75 | 82 | 90 | 96 | 111 | 97 |
| | 46.05 | 108.55 | 118.81 | 115.37 | 122.71 | 126.64 | 136.85 | 135.91 | 147.12 | 108.55 | 158.33 | 118.81 | 154.45 |
| | 38.10 | 97.83 | 119.20 | 113.94 | 138.37 | 135.31 | 166.96 | 159.74 | 191.39 | 184.17 | 227.56 | 220.34 | 215.82 |
| $x_4^{u} = 4$ | | | | | | (9) | (11) | (12) | | | | | |
| | 0.0150 | 0.1775 | 0.1455 | 0.0875 | 0.0785 | 0.0555 | 0.0491 | 0.0364 | 0.0300 | 0.0274 | 0.0271 | 0.0265 | 0.0210 |
| | 64 | 79 | 84 | 88 | 103 | 93 | 100 | 103 | 110 | 118 | 124 | 139 | 125 |
| | 60.47 | 122.97 | 133.23 | 129.79 | 137.13 | 140.06 | 151.27 | 150.33 | 161.54 | 157.67 | 172.75 | 165.67 | 168.88 |
| | 65.23 | 124.96 | 146.33 | 141.07 | 165.50 | 162.44 | 194.09 | 186.87 | 218.52 | 211.30 | 254.69 | 247.47 | 242.95 |

(Stage 4)

Table 10d    Computational results of Example 4-5 for Stage 1-2-3-4 and Stage 5

### Stage 1-2-3-4

| Number of components used | 2-2-1-3 | 3-2-1-3 | 2-2-2-3 | 2-2-3-3 | 3-2-2-3 | 4-2-2-3 | 3-3-2-3 | 4-3-2-3 | 3-2-2-4 | 3-3-3-3 | 4-2-2-4 | 3-3-2-4 |
|---|---|---|---|---|---|---|---|---|---|---|---|---|
| Stage unreliability | 0.2054 | 0.1734 | 0.1154 | 0.1064 | 0.0834 | 0.0770 | 0.0643 | 0.0579 | 0.0555 | 0.0553 | 0.0491 | 0.0364 |
| $g_1$ used | 51 | 56 | 60 | 75 | 65 | 72 | 75 | 82 | 93 | 90 | 100 | 103 |
| $g_2$ used | 108.55 | 118.81 | 115.37 | 122.71 | 126.64 | 136.85 | 185.91 | 147.12 | 140.06 | 108.55 | 151.27 | 150.33 |
| $g_3$ used | 97.83 | 119.20 | 113.94 | 138.37 | 135.31 | 166.96 | 159.74 | 191.39 | 162.44 | 184.17 | 194.09 | 186.87 |

### Stage 5

$x_5^{\ell} = 2$

| | (1) | (2) | (4) | (5) | (6) | | (8) | | (10) | | | |
|---|---|---|---|---|---|---|---|---|---|---|---|---|
| 0.0625 | 0.2679 | 0.2359 | 0.1779 | 0.1689 | 0.1459 | 0.1395 | 0.1268 | 0.1204 | 0.1180 | 0.1178 | 0.1116 | 0.0989 |
| 8 | 59 | 64 | 68 | 83 | 73 | 80 | 83 | 90 | 101 | 98 | 108 | 111 |
| 14.59 | 123.14 | 133.40 | 129.96 | 137.30 | 141.23 | 151.44 | 150.50 | 161.71 | 154.65 | 123.14 | 165.86 | 164.92 |
| 29.68 | 127.51 | 148.88 | 143.62 | 168.05 | 164.99 | 196.64 | 189.42 | 221.07 | 192.12 | 213.85 | 223.77 | 216.55 |

$x_5^{u} = 3$

| | (3) | | (7) | (9) | (11)* | | | | | | | |
|---|---|---|---|---|---|---|---|---|---|---|---|---|
| 0.0156 | 0.2210 | 0.1890 | 0.1310 | 0.1220 | 0.0990 | 0.0926 | 0.0799 | 0.0735 | 0.0711 | 0.0709 | 0.0647 | 0.0520 |
| 18 | 69 | 74 | 78 | 93 | 83 | 90 | 93 | 100 | 111 | 108 | 118 | 121 |
| 20.46 | 129.01 | 139.27 | 135.83 | 143.17 | 147.10 | 157.31 | 156.47 | 167.58 | 160.52 | 129.09 | 171.73 | 170.79 |
| 57.16 | 154.99 | 176.36 | 171.10 | 195.53 | 192.47 | 224.12 | 216.90 | 248.55 | 219.60 | 241.33 | 251.25 | 244.03 |

(3, 2).  Finally the dominating sequence of (1), (2), (3), (4), (5), (6), and (7) is obtained which is then the system composed of stages 1 and 2.

The dominating sequence for the combination of stage 1 and stage 2 from Table 10a will be the new row entry above the matrix in Table 10b. The number of components, stage unreliabilities, $g_1$, $g_2$, and $g_3$ of stage 3 will be the column to the left of the matrix of Table 10b. Similar procedures are now carried out to eliminate the entries of this matrix whose costs exceed the constraint, i.e., (4-4, 4); (4-4, 3); (5-3, 4); (5-3, 3); (4-3, 4).  The dominating sequence is then determined.  (5-3, 2) and (4-2, 4) are eliminated by comparing with (3-3, 3); (4-4, 1) and (5-3, 1) by (4-3, 2); (4-2, 3), (3-2, 4) and (3-2, 3) by (3-3, 2); (4-3, 1) by (4-2, 2); (3-3, 1) and (2-2, 4) by (3-2, 2); and (4-2, 1) by (2-2, 3).  The dominating sequence is (2-2, 1), (3-2, 1), (2-2, 2), (2-2, 3), (3-2, 2), (4-2, 2), (3-3, 2), (4-3, 2), (3-3, 3) (4-4, 2), (3-3, 4), and (4-3, 3).

The dominating sequence obtained for the system composed of stages 1, 2, and 3 then forms the row entries above the matrix of Table 10c. Stage 4 is combined with stages 1-2-3 to form a system, and its dominating sequence is obtained from Table 10c.  This dominating sequence for the system composed of stages 1, 2, 3, and 4 is used to combine with stage 5 to get the last dominating sequence as shown in Table 10d. In Table 10d, a dominating sequence is obtained, and the optimal one has the system configuration of (3, 2, 2, 3, 3) which has the highest reliability, $1 - 0.0990 = 0.9010$.


## REFERENCES


1. Bellman, R., *Dynamic Programming*, Princeton, N. J.: Princeton University Press (1957).

2. Bellman, R., and S. E. Dreyfus, *Applied Dynamic Programming*, Princeton, N. J.: Princeton University Press (1962).

3. Bellman, R., *Modern Analytic and Computational Methods in Science and Mathematics*, New York:  Elsevier (1968).

4.  Bellman, R. E., and S. E. Dreyfus, "Dynamic programming and re-
    liability of multicomponent devices," *Operations Research,* Vol.
    6, pp. 200-206 (1958).

5.  Black, G., and F. Proschan, "On optimal redundancy," *Operations
    Research,* Vol. 7, pp. 581-588 (1959).

6.  Burton, R. M., and G. T. Howard, "Optimal system reliability for
    a mixed series and parallel structure," *Journal of Mathematical
    Analysis and Applications,* Vol. 28, pp. 370-382 (1969).

7.  Fyffe, D. E., W. W. Hines, and N. K. Lee, "System reliability
    allocation and a computational algorithm," *IEEE Transactions on
    Reliability,* Vol. R-17, No. 2, pp. 64-69 (1968).

8.  Hadley, G., *Nonlinear and Dynamic Programming,* Reading, Mass:
    Addison-Wesley (1964).

9.  Jensen, P. A., "Optimization of series-parallel-series networks,"
    *Operations Research,* Vol. 18, pp. 471-482 (1970).

10. Kettelle, J. D., "Least-cost allocation of reliability investment,"
    *Operations Research,* Vol. 10, pp. 249-265 (1967).

11. Kulshrestha, D. K., and M. C. Gupta, "Use of dynamic programming
    for reliability engineers," *IEEE Transactions on Reliability,*
    Vol. R-22, pp. 240-241 (1973).

12. Lambert, B. K., A. G. Walvekar, and J. P. Hirmas, "Optimal re-
    dundancy and availability allocation in multistage systems,"
    *IEEE Transactions on Reliability,* Vol. R-20, No. 3, pp. 182-185
    (1971).

13. Liittschwager, J. M., "Dynamic programming in the solution of
    a multistage reliability problem," *Journal of Industrial Engi-
    neering,* Vol. 15-16, pp. 168-175 (1964-1965).

14. Messinger, M., and Shooman, H., "Technique for optimum spares
    allocation:  a tutorial review," *IEEE Transactions on Reliability,*
    Vol. R-19, pp. 156-166 (1970).

15. Nemhauser, G. L., *Introduction to Dynamic Programming,* New York:
    Wiley (1967).

16. Proschan, F., and T. A. Bray, "Optimum redundancy under multiple
    constraints," *Operations Research,* Vol. 13, pp. 800-814 (1965).

17. Rudd, D. F., "Reliability theory in chemical system design,"
    *I&EC Fundamental,* Vol. 1, No. 2, pp. 138-143 (1962).

18. Shershin, A. C., "Mathematical optimization techniques for the
    simultaneous apportionments of reliability and maintainability,"
    *Operations Research,* Vol. 18, No. 1, pp. 95-106 (1970).

19.  Tillman, F. A., and J. M. Liittschwager, "Integer programming
     formulation of constrained reliability problems," *Management
     Science,* Vol. 13, No. 11, pp. 887-899 (1967).

20.  Woodhouse, C. F., "Optimal redundancy allocation by dyanmic pro-
     gramming," *IEEE Transactions on Reliability,* Vol. R-21, No. 1,
     pp. 60-62 (1972).

## 5.1    INTRODUCTION

A simple computational procedure based on the discrete maximum principle has been developed for maximizing reliability of multistage parallel systems subject to multiple nonlinear constraints [1]. It appears that the procedure can be applied to a variety of optimization
problems with separable and multiple constraints functions.

## 5.2    STATEMENT OF THE PROBLEM AND THE COMPUTATIONAL PROCEDURE

The problem of maximizing the reliability of an N-stage series
system with redundant units in parallel (see Fig. 1) subject to multiple linear and nonlinear separable constraints can be stated as follows:
Maximize

$$R_s = \prod_{n=1}^{N} \left( 1 - (1 - R^n)^{\theta^n} \right) \qquad\qquad (1)\,\dagger$$

---

$\dagger$ The superscript n indicates the stage number. The exponents are
written with parentheses or brackets such as $(x^n)^2$ or $\{T^n(x^{m-1};\theta^n)\}^2$.

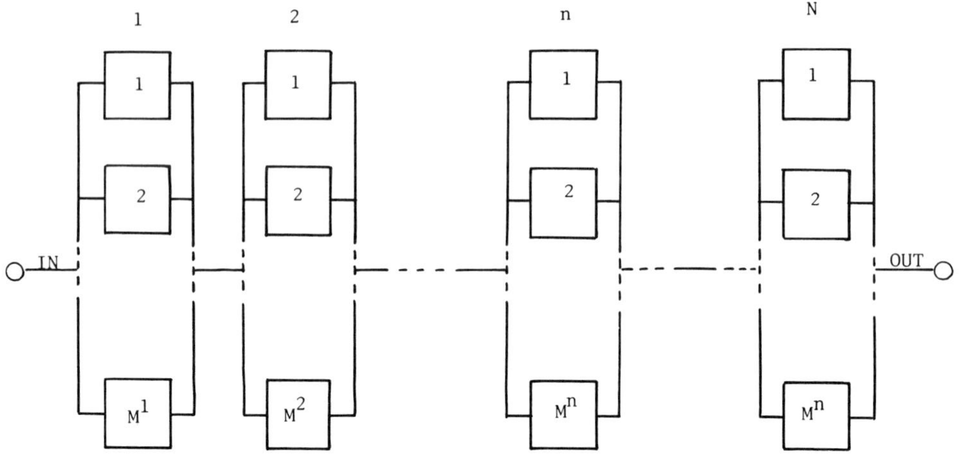

Figure 1    A mixed system with N-stages in series where components
are in parallel at each stage.


subject to

$$\sum_{n=1}^{N} g_i^n(\theta^n) \leq b_i, \qquad i = 1,2,\ldots,s \tag{2}$$


where $R_s$ is the system reliability, N, the total number of stages,
$R^n$, the reliability of one element at the $n^{th}$ stage, $\theta^n$, the number of
elements at the $n^{th}$ stage, where $(\theta^n-1)$ is the number of redundant
units, $g_i(\theta^n)$, the function representing the amount of the $i^{th}$ resource consumed at the $n^{th}$ stage as a function of $\theta^n$, s, the number
of constraints, and $b_i$, the total amount of the $i^{th}$ resource available.
Let $x_i^n$ = the $i^{th}$ resource corresponding to the $i^{th}$ constraint, which
is consumed in the first n stages, $i = 1,2,\ldots,s$.  Then, the performance equations for this N-stage system may be written as

$$x_i^n = x_i^{n-1} + g_i^n(\theta^n), \qquad n = 1,2,\ldots,N \tag{3}$$
$$i = 1,2,\ldots,s$$

$$x_i^0 = 0 \tag{3a}$$

$$x_i^N \leq b_i \tag{3b}$$

By defining

$$x_{s+1}^n = x_{x+1}^{n-1} + \ell n \ [1 - (1 - R^n)^{\theta^n}], \qquad n = 1,2,\ldots,N \tag{4}$$

$$x_{s+1}^0 = 0$$

the objective function to be optimized can be written as

$$S = \ell n \ R_s$$

$$= x_{s+1}^N$$

$$= \sum_{i=1}^{s+1} c_i x_i^N \tag{5}$$

where

$$c_i = 0, \qquad i = 1,2,\ldots,s \tag{6}$$

$$c_{s+1} = 1$$

The Hamiltonian and the adjoint variables of the system can be defined as (see Appendix A2)

$$H^n = \sum_{i=1}^{s+1} z_i^n x_i^n$$

$$= \sum_{i=1}^{s} z_i^n (x^{n-1} + g_i^n(\theta^n)) + z_{s+1}^n (x_{s+1}^{n-1} + \ell n \ (1 - (1 - R^n)^{\theta^n}) \tag{7}$$

$$n = 1,2,\ldots,N$$

$$z_i^{n-1} = \frac{\partial H^n}{\partial x_i^{n-1}} = z_i^n, \qquad n = 1, 2, \ldots, N \tag{8}$$

$$i = 1, 2, \ldots, s, \; s+1$$

$$z_{s+1}^N = c_{s+1} = 1 \tag{9}$$

Equations (8) and (9) yield

$$z_{s+1}^n = 1, \qquad n = 1, 2, \ldots, N \tag{10}$$

Assuming that the non-trivial and unique Hamiltonian and adjoint variables of the system exist, the necessary stationary condition for local optimality can be obtained as

$$\frac{\partial H^n}{\partial \theta^n} = 0 = \sum_{i=1}^{s} z_i^n \frac{\partial g_i^n(\theta^n)}{\partial \theta^n} + \frac{-(1 - R^n)^{\theta^n} \ln (1 - R^n)}{1 - (1 - R^n)^{\theta^n}} \tag{11}$$

In employing this condition in determing the optimal condition of the system, we assume its existence. In reality, $\theta^n$, $n = 1, 2, \ldots, N$ are positive integers. We assume, however, that $\theta^n$ are continuous variables.

Now we assume that one of the constraints, say the $j^{th}$ constraint given by eq. (2) or equivalently by eq. (3), is active and the rest are free. This means the end condition corresponding to the $j^{th}$ constraint is fixed and the rest of them are free. Then we have

$$z_i^n = c_i = 0, \qquad i = 1, 2, \ldots, s \tag{12}$$

$$i \neq j$$

From eqs. (8) and (12), we obtain

$$z_i^n = 0, \qquad i = 1, 2, \ldots, s$$

$$i \neq j$$

$$n = 1, 2, \ldots, N$$

Therefore, eq. (11) reduces to

$$z_j^n \, \frac{\partial g_j^n(\theta^n)}{\partial \theta^n} \; - \; \frac{(1 - R^n)^{\theta^n} \, \ln \, (1 - R^n)}{1 - (1 - R^n)^{\theta^n}} = 0 \tag{13}$$

The procedure for solving the problem may be described in the following steps.

Step 1.  Assuming a value for $\theta^1$ in eq. (13), we obtain $z_j^1$. Furthermore, eq. (8) gives

$$z_j^1 = z_j^n, \qquad n = 2,\ldots,N$$

Step 2.   We find $\theta^n$, $n = 2,3,\ldots,N$ from eq. (13) by using the values of $z_j^n$ obtained.

Step 3.   We compute $x_i^N$, $i = 1,2,\ldots,s$ from eq. (3).

Step 4.   One of the following conditions will occur.

a)  If $x_i^N < b_i$ for all $i = 1,2,\ldots,s$ then we assume a higher value for $\theta^1$ and return to step 1.

b)  If $x_j^N > b_j$ and $x_{i_1}^N < b_i$ for $i \neq j$, $i = 1,2,\ldots,s$, we assume a smaller value for $\theta^1$ and return to step 1.

c)  If $x_k^N > b_k$, $k \neq j$ and $x_i^N < b_i$, $i = 1,2,\ldots,j,\ldots,s$, $i \neq k$, where $j$ is the active constraint, then we go to step 5.

d)  If $x_j^N = b_j$, and $x_i^N < b_i$, $i = 1,2,\ldots,s$, $i \neq j$, that is, the $j^{th}$ constraint reaches its limit while none of the other constraints are violated, we have a candidate for the optimal solution.

Step 5.   We replace constraint j by constraint k.  Accordingly we replace j by k in eq. (13) and in steps 1 and 2 and repeat the procedure given by step 1 through step 4.

## 5.3   EXAMPLE

Constraints on a system may be the total weight, the total cost, the total volume and so on. In general, such constraints are nonlinear.

As the number of units at each stage is increased, the number of connecting components must also increase, and thus, the cost and weight may increase exponentially.

Let

$c^n$ = cost per element at the $n$th stage,

$w^n$ = weight per element at the $n$th stage,

$v^n$ = volume per element at the $n$th stage,

$\theta^n$ = number of elements in parallel at the $n$th stage.

Therefore the following nonlinear constraints for the combination of weight and volume, cost, and weight are considered.

(1)   The constraint which is imposed for the combination of weight and volume is

$$\sum_{n=1}^{N} g_1^n(\theta^n) = \sum_{n=1}^{N} p^n(\theta^n)^2 \leq P$$

where $p^n = w^n v^n$ is the product of weight per unit and volume per unit at the $n$th stage.

(2)   The cost constraint is

$$\sum_{n=1}^{N} g_2^n(\theta^n) = \sum_{n=1}^{N} c^n(\theta^n + \exp(\theta^n/4)) \leq C$$

where $c^n \theta^n$ is the cost of units at the $n$th stage and $c^n(e)^{\theta^n/4}$ is the additional cost for interconnecting parallel units [4].

(3)   The weight constraint is

$$\sum_{n=1}^{N} g_3^n(\theta^n) = \sum_{n=1}^{N} w^n \theta^n \exp(\theta^n/4) \leq W$$

where $w^n \theta^n$ is the weight of the total units at the $n$th stage.  This is increased by the factor $\exp(\theta^n/4)$ due to the weight of the interconnecting links [4].

The problem is to maximize the system reliability subject to the above constraints.  State variables of the system are defined as follows:

$$x_1^n = x_1^{n-1} + p^n (\theta^n)^2, \qquad n = 1, 2, \ldots, N \tag{14}$$

$$x_1^0 = 0$$

$$x_1^N \leq P$$

$$x_2^n = x_2^{n-1} + c^n \left[ \theta^n + (e)^{\theta^n/4} \right], \qquad n = 1, 2, \ldots, N \tag{15}$$

$$x_2^0 = 0$$

$$x_2^N \leq C$$

$$x_3^n = x_3^{n-1} + w^n \theta^n (e)^{\theta^n/4}, \qquad n = 1, 2, \ldots, N \tag{16}$$

$$x_3^0 = 0$$

$$x_3^N \leq W$$

$$x_4^n = x_4^{n-1} + \ln\{1 - (1 - R^n)^{\theta^n}\}, \qquad n = 1, 2, \ldots, N \tag{17}$$

$$x_4^0 = 0$$

The objective function to be maximized is

$$S = \sum_{n=1}^{N} \ln \{1 - (1 - R^n)^{\theta^n}\}$$

$$= \sum_{i=1}^{4} c_i x_i^N \tag{18}$$

$$= x_4^N$$

where

$$c_i = 0, \qquad i = 1,2,3 \tag{19}$$

$$c_4 = 1$$

The Hamiltonian and adjoint variables of the system are

$$H^n = \sum_{i=1}^{4} z_i^n x_i^n$$

$$= z_1^n (x_1^{n-1} + p^n (\theta^n)^2) + z_2^n \{ x_2^{n-1} + c^n (\theta^n + (e)^{\theta^n}/4) \}$$

$$+ z_3^n \left[ x_3^{n-1} + w^n \theta^n (e)^{\theta^n}/4 \right] + z_4^n \left[ x_4^{n-1} + \ln\{1 - (1 - R^n)^{\theta^n}\} \right] \tag{20}$$

$$n = 1,2,\ldots,N$$

$$z_i^{n-1} = \frac{\partial H^n}{\partial x_i^{n-1}} = z_i^n, \qquad i = 1,2,3,4 \tag{21}$$

$$z_4^N = c_4 = 1 \tag{22}$$

From eqs. (21) and (22), we obtain

$$z_4^N = 1, \qquad n = 1,2,\ldots,N$$

Differentiating eq. (20) with respect to $\theta^n$ and equating to zero, we obtain

$$\frac{\partial H^n}{\partial \theta^n} = 0$$

$$= 2z_1^n p^n \theta^n + z_2^n c^n \left[ 1 + \frac{1}{4} (e)^{\theta^n}/4 \right]$$

$$+ z_3^n w^n \left[ (e)^{\theta^n/4} + \frac{1}{4} \theta^n (e)^{\theta^n/4} \right] + \frac{- (1 - R^n)^{\theta^n} \ln (1 - R^n)}{1 - (1 - R^n)^{\theta^n}} \qquad (23)$$

Whenever the $j^{th}$ constraint, represented by $x_j^N$, is active, this has the effect of fixing its boundary value.  Thus

$$z_i^N = c_i, \qquad i \neq j \qquad (24)$$

Now, if the first constraint is the only one active, we obtain the following relations from eqs. (24), (21), and (19).

$$z_i^N = c_i = 0, \qquad i = 2,3$$

and

$$z_i^n = 0, \qquad n = 1,2,\ldots,N$$
$$i = 2,3$$

Consequently eq. (23) can be written as

$$2z_1^n p^n \theta^n = \frac{(U^n)^{\theta^n} \ln U^n}{1 - (U^n)^{\theta^n}} \qquad (25)$$

where

$$U^n = 1 - R^n$$

Rearranging the terms in eq. (25), we have

$$z_1^n = \frac{1}{2p^n \theta^n} \cdot \frac{(U^n)^{\theta^n} \ln U^n}{1 - (U^n)^{\theta^n}} \qquad (26)$$

and

$$\theta^n = (U^n)^{\theta^n} \left[ \theta^n + \frac{\ln U^n}{2z_1^n p^n} \right] \tag{27}$$

Letting

$$A^n = \frac{\ln U^n}{2z_1^n p^n}$$

eq. (27) becomes

$$\theta^n = (U^n)^{\theta^n} (\theta^n + A^n)$$

or

$$f(\theta^n) = \theta^n - (U^n)^{\theta^n} (\theta^n + A^n) = 0 \tag{28}$$

This equation can be solved by Newton's method for $\theta^n$.

   Similarly, if the second constraint is active and the rest are free, we obtain the following relations.

$$z_i^N = c_i = 0, \qquad i = 1,3$$

$$z_i^n = 0, \qquad n = 1,2,\ldots,N$$

$$i = 1,3$$

$$z_2^n c^n \left[ 1 + \frac{1}{4}(e)^{\theta^n/4} \right] = \frac{(U^n)^{\theta^n} \ln U^n}{1 - (U^n)^{\theta^n}} \tag{29}$$

$$z_2^n = \frac{1}{c^n \left( 1 + \frac{1}{4}(e)^{\theta^n/4} \right)} \left[ \frac{(U^n)^{\theta^n} \ln U^n}{1 - (U^n)^{\theta^n}} \right] \tag{30}$$

and

$$f(\theta^n) = \left[ 1 + \frac{1}{4}(e)^{\theta^n/4} \right] - (U^n)^{\theta^n} \left[ (1 + \frac{1}{4}(e)^{\theta^n/4}) + \frac{\ln U^n}{z_2^n c^n} \right] = 0 \tag{31}$$

This equation is well behaved and Newton's method can be employed to obtain $\theta^n$.

Similarly, if the third constraint is active and the rest are free we obtain the following relations.

$$z_i^N = c_i = 0, \qquad i = 1,2$$

$$z_i^n = 0, \qquad n = 1,2,\ldots,N$$

$$i = 1,2$$

$$z_3^n w^n \left( (e)^{\theta^n/4} \right) = \frac{(U^n)^{\theta^n} \ln U^n}{1 - (U^n)^{\theta^n}} \tag{32}$$

$$z_3^n = \frac{1}{w^n \left( (e)^{\theta^n/4} + \frac{1}{4} \theta^n (e)^{\theta^n/4} \right)} \left[ \frac{(U^n)^{\theta^n} \ln U^n}{1 - (U^n)^{\theta^n}} \right] \tag{33}$$

$$f(\theta^n) = (e)^{\theta^n/4} (1 + \frac{1}{4} \theta^n) - (U^n)^{\theta^n} \left\{ (e)^{\theta^n/4}(1 + \frac{1}{4} \theta^n) \right.$$

$$\left. + \frac{\ln U^n}{z_3^n w^n} \right\} = 0 \tag{34}$$

This is again a well behaved equation and can be solved by Newton's method.

## 5.4   NUMERICAL RESULTS

A five stage problem was solved with the constants given in Table
1.   The optimum redundancy obtained is as follows.

$$\theta^1 = 2.6000$$
$$\theta^2 = 2.2816$$
$$\theta^3 = 2.0075$$
$$\theta^4 = 2.6882$$
$$\theta^5 = 3.3981$$

Since $\theta^n$, $n = 1,2,\ldots,5$ in reality, should be positive integers, we
approximately obtain

$$\theta^1 = 3$$
$$\theta^2 = 2$$
$$\theta^3 = 2$$
$$\theta^4 = 3$$
$$\theta^5 = 3$$

The number of redundant elements at each stage can be obtained by sub-
tracting one from each of the above figures.

From the result we find that the total of the product of weight
and volume is 83 with a slack of 27 units, the cost of the system is
146.12 with a slack of 28.88 units and the weight of the system is
192.48 with a slack of 7.52 units. This policy results in a system

Table 1    Constants assigned for 5 stage problem.

| $n$ | $R^n$ | $p^n$ | $P$ | $c^n$ | $C$ | $w^n$ | $W$ |
|---|---|---|---|---|---|---|---|
| 1 | .80 | 1 |     | 7 |     | 7 |     |
| 2 | .85 | 2 |     | 7 |     | 8 |     |
| 3 | .90 | 3 | 110 | 5 | 175 | 8 | 200 |
| 4 | .65 | 4 |     | 9 |     | 6 |     |
| 5 | .75 | 2 |     | 4 |     | 9 |     |

reliability of 0.9045.  A numerical simulation indicated that the above result is the optimum.


## 5.5    CONCLUSION

A simple and practical computational procedure is presented for maximizing the reliability of a system under multiple nonlinear constraints.  An example with three nonlinear constraints is solved in detail to illustrate the method.  Problems with multiple linear constraints are special cases of the problems presented here.

The objective function given by eq. (4), that is, the logarithm of the system reliablity given by eq. (1), and the functions representing the constraints given by eq. (2), are separable functions.  Therefore, the present method may be applied in general to optimization of problems with separable objective and constraint functions.

In applying the above technique it is assumed that the optimal sequence of $\theta^n$ is obtained by using the recurrence relation given by eq. (13) when only the $j^{th}$ constraint is active.  In the computational procedure only the necessary condition for local optimality is used to obtain the candidate for the optimal solution.  Therefore, simulation is involved to assume numerically the sufficiency of the optimal solution.  In spite of this shortcoming the present method appears to overcome some of the practical limitations of other methods used to solve this class of problems.

The content of this chapter is based on Tillman et al. [5].


## REFERENCES

1.    Fan, L. T., C. S. Wang, F. A. Tillman, and C. L. Hwang, Optimization of Systems Reliability," *IEEE Transactions on Reliability*, vol. R-16, pp. 81-86 (1967).

2.    Fan, L. T., and C. S. Wang, *The Discrete Maximum Principle - A Study of Multistage Systems Optimization*, New York: Wiley, (1964).

3.  Fan, L. T., *The Continuous Maximum Principle - A Study of Complex Systems Optimization,* New York: Wiley (1966).

4.  Tillman, F. A., and J. Liittschwager, "Integer Programming Formulation of Constrained Reliability Problems," *Management Science,* vol. 13, pp 887-899 (1967).

5.  Tillman, F. A., C. L. Hwang, L. T. Fan, and S. A. Balbale, "Systems Reliability Subject to Multiple Nonlinear Constraints," *IEEE Transaction on Reliability,* vol. R-17, No. 3, pp. 153-157 (1968).

# CHAPTER 6    SEQUENTIAL UNCONSTRAINED MINIMIZATION TECHNIQUE (SUMT) APPLIED TO OPTIMAL SYSTEMS RELIABILITY

## 6.1    INTRODUCTION

The problems considered in this section involve optimization of systems reliability of a complex system. The optimization method employed is the Sequential Unconstrained Minimization Technique (SUMT). This method is considered one of the simplest and most efficient methods for solving constrained nonlinear programming problems.

The principle of SUMT is a transformation of a constrained minimization problem into a sequence of unconstrained minimization problems. This transformation enables us to use well-established unconstrained optimization techniques to solve the constrained problem without inventing a new technique. The method was first proposed by Carroll in 1959 [1,2] and further developed by Fiacco and McCormick [3,4,5,6,12]. In 1964, Fiacco and McCormick developed a general algorithm based on SUMT, and in 1965, they proposed a method which is called SUMT without parameters. By using this method the difficulty of choosing the penalty parameters can be avoided, although some minor difficulties may still exist. There is a general computer program provided by McCormick, Mylander and Fiacco called "RAC Computer Program Implementing the Sequential Unconstrained Minimization Technique for Nonlinear Programming," (IBM SHARE number 3189) [12]. In this computer program, the unconstrained minimization technique used is the second order gradient method.

Difficulties which arise from use of the second order gradient method as an unconstrained minimization technique in SUMT become

predominant in a large size and/or very complex nonlinear problem.
The difficulties arise particularly in obtaining the first and second
order partial derivatives of very complex nonlinear functions which
most practical problems have.  Therefore, a new algorithm which uses
a much simpler direct search technique is needed.

For these reasons, a new technique of implementing SUMT with
the Hooke and Jeeves pattern search technique as its unconstrained
minimization process is suggested.  The procedures are presented in
[9,11].  The Hooke and Jeeves pattern search technique [7,8] is different from the gradient method in deciding the direction of search.
In the gradient method it is the direction of the steepest descent
while that of the Hooke and Jeeves pattern search technique is determined by a direct comparison with the values of the objective function
at two separate points a finite step apart.  For this reason, when
the pattern search is getting close to the boundary of some inequality
constraints, it will frequently go out of the feasible region bounded
by these constraints, and the search might be terminated at some point
near the boundary which may not be the optimum.  To avoid this, heuristic programming technique developed by Paviani and Himmelblau [13]
is used; it enables one to make turns at the pattern search near the
boundary of constraints and continue. The details of the method are
described in [9,11] along with a general FORTRAN-IV program and the
computer diagrams.

The optimization of the complex system reliability problem using
the RAC-SUMT computer program has been carried out in [11,14], and
also by using the KSU-SUMT computer program in [9,11].  In this chapter, the same complex system problems with a better cost function [15]
were solved with the KSU-SUMT program.


## 6.2    FORMULATION OF THE PROBLEM


A system whose redundant units are not in a purely series configuration is considerably more difficult to solve.  One such example
is shown in Fig. 1.  In this system, unit 1 is backed up by a parallel

unit 4. There are two equal paths, each of which has unit 2 in series with the stage formed by units 1 and 4. These two equal paths operate in parallel so that if one of them is good the output is assured. However, because unit 2 does not have a high degree of reliablity, a third unit, unit 3, is inserted into the circuit. Therefore, the following operations are possible: 2-1, 2-4, 3-1, and 3-4, and each operation has equal paths.

In attempting to optimize the reliability of a system with such a configuration, a major difficulty is encountered in that the reliability expression is not a separable function and thus cannot be analyzed as a multistage process. Hence a different approach is used to solve this problem: Bayes' theorem is used, which utilizes conditional probabilities [16], to obtain the nonlinear system reliability which is subject to some constraints. This nonlinear programming problem is then solved using SUMT.

## SYSTEM RELIABILITY USING CONDITIONAL PROBABILITIES

In solving this problem, a simplified form of Bayes' probability theorem is used. The theorem states that if A is an event that depends on one or two mutually exclusive events $B_i$ and $B_j$ of which one must necessarily occur, then the probability of the occurrence of A is given by

$$P(A) = P(A, \text{ given } B_i)P(B_i) + P(A, \text{ given } B_j)P(B_j) \tag{1}$$

Let $Q_s$ represent the probability of system failure, $R_k$ the probability that component K is good, and $Q_k$ the probability that component K is bad. We then obtain the following expression for system unreliability.

$$Q_s = Q_s(\text{given } K \text{ is good})R_k + Q_s(\text{given } K \text{ is bad})Q_k \tag{2}$$

The corresponding system reliability $R_s$ is

$$R_s = 1 - Q_s \tag{3}$$

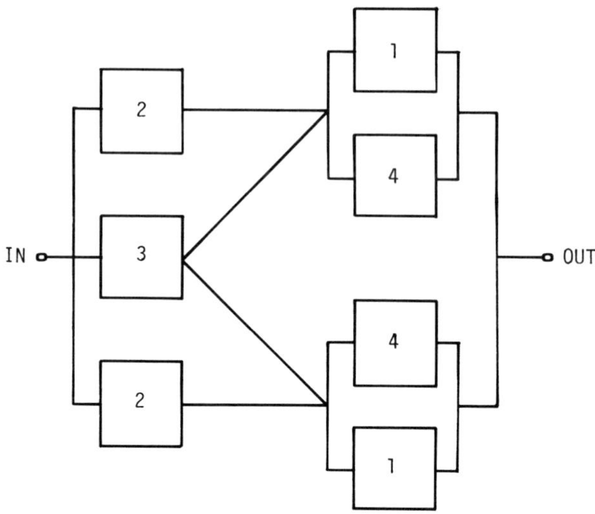



Figure 1     A schematic diagram of a complex system.

To obtain the reliability of the system presented in Fig. 1 we select component 3 as the key component in eq. (2), denoted by K. Thus we have the expression for the system unreliablity

$$Q_s = Q_s(\text{if 3 is good})R_3 + Q_s(\text{if 3 is bad})Q_3 \qquad (4)$$

If component 3 is good, the system can fail if the stage formed by units 1 and 4 fails.  Thus, the system's unreliability, given that unit 3 is good, is

$$Q_s(\text{if 3 is good}) = [(1 - R_1)(1 - R_4)]^2 \qquad (5)$$

If, on the other hand, unit 3 is bad the system's unreliability is

$$Q_s(\text{if 3 is bad}) = \{1 - R_2[1 - (1 - R_1)(1 - R_4)]\}^2 \qquad (6)$$

From (4) the unreliability of the system is

$$Q_s = [(1 - R_1)(1 - R_4)]^2 R_3$$

$$+ \{1 - R_2[1 - (1 - R_1)(1 - R_4)]\}^2(1 - R_3) \tag{7}$$

The assumption is made that the reliablity of the components are
independent of each other.  That is, the reliability of component 4
would not be affected by the failure of component 1.  The system reliability is then given by eq. (3).


## 6.3    COMPUTATIONAL PROCEDURES OF SUMT

The general nonlinear programming problem with nonlinear inequality and/or equality constraints is one of choosing $x$ so as to
minimize

$$f(x)$$

subject to

$$g_i(x) \geq 0, \qquad i = 1,\ldots,m \tag{8}$$

$$h_j(x) = 0, \qquad j = 1,\ldots,\ell$$

The SUMT technique for solving eq. (8) is based on minimizing the
function

$$P(x, r_k) = f(x) + r_k \sum_{i=1}^{m} [g_i(x)]^{-1} + r_k^{-\frac{1}{2}} \sum_{j=1}^{\ell} h_j^2(x) \tag{9}$$

over a strictly monotonic decreasing sequence $\{r_k\}$.  $P(x, r_k)$ is minimized with respect to $x(r_k)$ for a given value of $r_k$.  The sequence
of values of $\{P(x, r_k)\}$ converges to the constrained optimum value
of the original objective function, $f(x)$, as $\{r_k\} \to 0$.  The essential
requirement is the convexity of the P-function.  Mathematical proof
of the convergence of the method is given in [6].

A new heuristic programming algorithm for using SUMT which incorporates the Hooke and Jeeves pattern search is summarized below [9, 11].

Step 1. Select a starting point $x^0$, the initial value of the penalty coefficient $r^0$, the initial tolerance limit for violating the constraints $B^0$, and the initial step-sizes $d^0$ needed in the search procedure.

Step 2. Go to step 3 if $x^0$ is feasible (viz., inside the region bounded by the inequality constraints). Otherwise select a feasible starting point by minimizing the total weight of the violating constrtraints. This weight, TGH, is defined by

$$(\text{TGH})^2 \equiv \sum_{t\varepsilon T} g_t^2(x^0) + \sum_{s\varepsilon S} h_s^2(x^0)$$

where $T \equiv \{t | g_t(x) < 0\}$, $S \equiv \{s | h_s(x) \neq 0\}$. TGH includes only the violated constraints.

Step 3. Minimize P in eq. (9) by the Hooke and Jeeves pattern search technique. Check after every move: if the move goes outside the feasible region, go to step 4; otherwise, if x* is reached for the current $r_k$, go to step 5.

Step 4. Move back to the near-feasible region and then return to step 3. The near-feasible region is defined as the region where all points satisfy the condition: TGH < B, where B is the tolerance limit of the violating constraint sequentially decreased after every violation to the inequality constraints.

Step 5. Check to see if the x* obtained in step 3 is feasible. If x* is feasible, go to step 7; otherwise go to step 6.

Step 6. Move x* (in the infeasible region) back into the feasible region along the direction toward the last optimum point; then go to step 7.

Step 7. Check to see if the stopping criterion, such as

$$\left\| \left| \frac{f(x^*)}{G(x^*, r_k)} \right| - 1 \right| < \varepsilon \tag{10}$$

is satisfied. The solution is then the optimal one if the criterion is satisfied; otherwise, go to step 8.  $G(x, r_k)$ in eq. (10) is defined as [6]

$$G(x, r_k) \equiv f(x) - r_k \sum_{j=1}^{m} [g_i(x)]^{-1} + r_k^{-1/2} \sum_{j=1}^{\ell} h_j^2(x) \tag{11}$$

Step 8.  Set $k \leftarrow k + 1$; $r_{k+1} \leftarrow r_k/Z$, where $Z$ is a constant greater than 1; and $d^{k+1} \leftarrow d^0/(k + 1)$.  Return to step 3.

The flow diagram is shown in Fig. 2.  The detailed discussions about the "procedure for selecting a feasible starting point from the infeasible initial point", the "computational procedure for minimizing $P(\bar{x}, r_k)$ function by the Hooke and Jeeves pattern search", the "procedure for moving an infeasible point into the feasible or near-feasible region bounded by inequality constraints", and the "procedure for moving the near-feasible k-th sub-optimum into the feasible region" are all referenced in Lai [11].

## 6.4  NUMERICAL EXAMPLES

*Example 6-1.*  The problem of maximizing the reliability of the complex system given in Fig. 1, which is subject to a single constraint can be stated as follows using eqs. (3) and (7).

Maximize the system reliability

$$R_s = 1 - Q_s$$

$$= 1 - R_3[(1 - R_1)(1 - R_4)]^2$$

$$-(1 - R_3)\{1 - R_2[1 - (1 - R_1)(1 - R_4)]\}^2 \tag{12}$$

subject to

$$C_s = \sum_i C_i \leq C$$

$$R_i \geq R_{i,min} \tag{13}$$

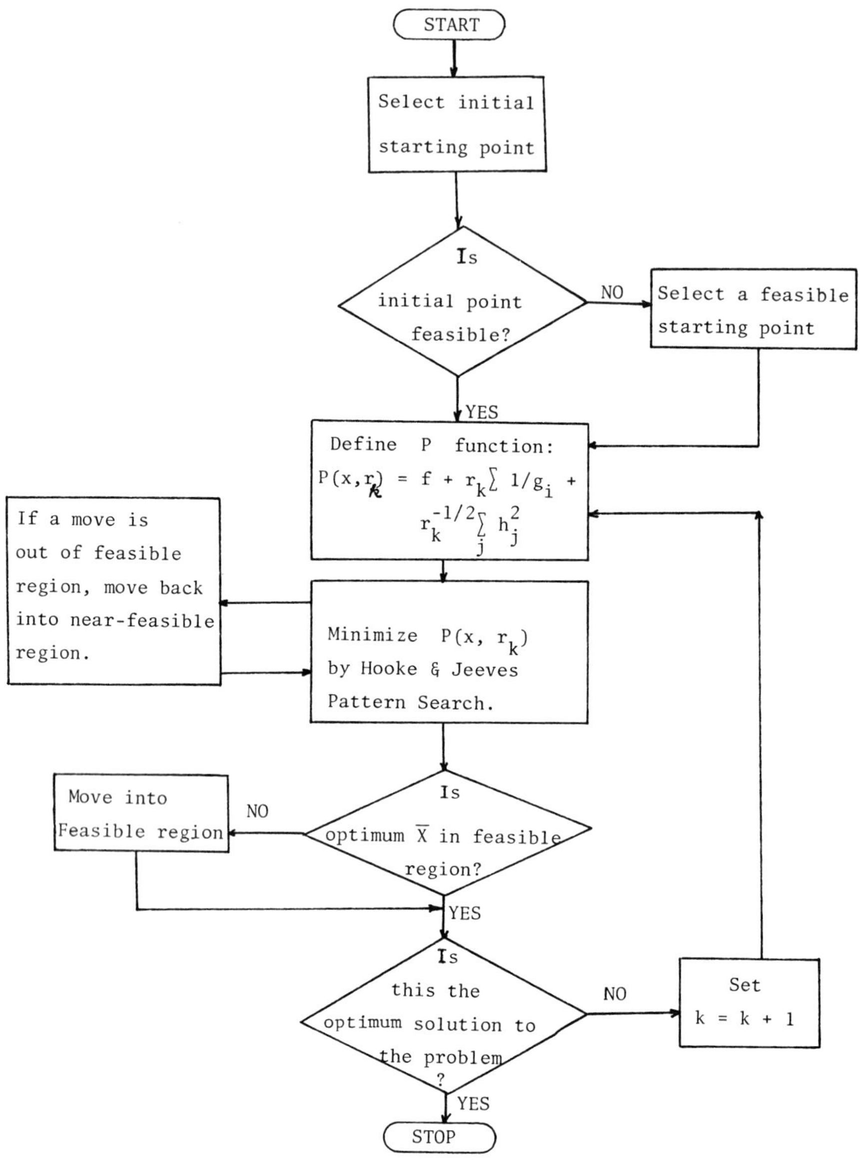


Figure 2  Descriptive flow diagram for SUMT with Hooke and Jeeves pattern search.

where

$$C_i \equiv K_i R_i^{\alpha_i} \tag{14}$$

The constraint given by eq. (13) can be interpreted as follows. $C_i$ can represent the weight, cost, or volume of each unit or component of the system, and the total weight, cost, or volume of the system must be less than C. Each of these is a function of reliability that can be expressed by eq. (14) where $K_i$ is a proportionality constant and $\alpha_i$ the exponential factor that relates $C_i$ and the reliability. That is, $K_i$ is the weight, cost, or volume of the component when

$R \equiv 1$ and $K_i R_i^{\alpha_i}$ is the reduced cost, weight, or volume when $R_i < 1$. Usually $\alpha_i$ is less than one. The following values are assigned to the constants $K_1$, $K_2$, $K_3$, and $K_4$, the constraint C and the exponential constant $\alpha_i$, $i = 1,2,3,4$.

$$K_1 = 100, \qquad K_2 = 100, \qquad K_3 = 200, \qquad K_4 = 150,$$

$$C = 800, \qquad \alpha_i = 0.6, \qquad i = 1,2,3,4.$$

The problem is formulated in a format to be solved by SUMT as follows.

Minimize

$$f(x) = -R_s$$

$$= -1 + R_3[(1 - R_1)(1 - R_4)]^2$$
$$+ (1 - R_3)\{1 - R_2[1 - (1 - R_1)(1 - R_4)]\}^2$$

subject to the constraints

$$g_1(x) = C - (2K_1 R_1^{\alpha_1} + 2K_2 R_2^{\alpha_2} + K_3 R_3^{\alpha_3} + 2K_4 R_4^{\alpha_4}) \geq 0$$

$$g_{i+1}(x) = 1 - R_i \geq 0, \qquad i = 1,2,3,4$$

$$g_{i+5}(x) = R_i - R_{i,min} \geq 0, \qquad i = 1,2,3,4$$

The P-function of eq. (9) is

$$P(x, r_k) = - 1 + R_3 [(1 - R_1)(1 - R_4)]^2$$

$$+ (1 - R_3) \{1 - R_2 [1 - (1 - R_1)(1 - R_4)]\}^2$$

$$+ r_k \left\{ \frac{1}{\left[ C - (2K_1^{\alpha_1} + 2K_2 R_2^{\alpha_2} + K_3 R_3^{\alpha_3} + 2K_4 R_4^{\alpha_4}) \right]} \right.$$

$$\left. + \sum_{i=1}^{4} \left[ \frac{1}{1 - R_i} + \frac{1}{R_i - R_{i,min}} \right] \right\}$$

The solutions obtained from two sets of initial reliabilities
for components, namely, $[R_1, R_2, R_3, R_4] = [0.7, 0.7, 0.7, 0.7]$ and
$[R_1, R_2, R_3, R_4] = [0.6, 0.6, 0.6, 0.6]$, are presented in Table 1
together with the corresponding results obtained by RAC program [12].
The solutions are almost identical, that is, the optimal system reliability, $R_s$, of 0.999998 is obtained with the cost of 799.733 for
the first set of starting components reliabilities, and the optimal
system reliability, $R_s$, of 0.999997 is obtained with the cost of
799.908 for the second set of starting components reliabilities. Recall that the constraint on the cost is 800. The component reliabilities are almost the same for both sets of starting points. The stopping criterion for terminating the minimization of the P-function
at each k iteration is when the number of cut-down step-size operations
in the Hooke and Jeeves pattern search is 3, and the final stopping
criterion for terminating the problem is $\varepsilon = 10^{-4}$. For the first
set of starting points, it takes 12 iterations for the P-functions,
$k = 12$, with a total of 1192 f-functional values evaluated. And for
the second set, 12 iterations for the P-functions, $k = 12$, with a
total of 1194 f-functional values evaluated.

Tables 2a and 2b present the iteration results for the solution
converging toward the optimal solution. Results given in these tables
show that the system reliability $R_s$, is monotonically increasing as

Table 1   Comparison of the Solutions of the System Reliability Maximization Problem (Example 6-1)

| Program | Number of k iterated | Component | reliability | | | System reliability | Cost | Stopping criteria for each k | for final | Computing time |
|---|---|---|---|---|---|---|---|---|---|---|
| | | $R_1$ | $R_2$ | $R_3$ | $R_4$ | | | | | |
| RAC | 0 | 0.7 | 0.7 | 0.7 | 0.7 | | | $\varepsilon'=10^{-5}$ | $\varepsilon=10^{-4}$ | exceeds 20 min. |
| | 10 | 0.9876 | 0.9936 | 0.6972 | 0.6941 | 0.99996 | 799.78 | | | |
| program | 0 | 0.6 | 0.6 | 0.6 | 0.6 | | | | | exceeds 20 min. |
| | 11 | 0.9889 | 0.9921 | 0.7019 | 0.6886 | 0.99995 | 799.28 | $\varepsilon'=10^{-5}$ | $\varepsilon=10^{-4}$ | |
| new | 0 | 0.7 | 0.7 | 0.7 | 0.7 | | | | | (both problems together) |
| | 12 | 0.997626 | 0.998399 | 0.682652 | 0.694958 | 0.999998 | 799.733 | INCUT=3 | $\varepsilon=10^{-5}$ | |
| program | 0 | 0.6 | 0.6 | 0.6 | 0.6 | | | | | |
| | 12 | 0.997409 | 0.998117 | 0.702590 | 0.68217 | 0.999997 | 799.908 | INCUT=3 | $\varepsilon=10^{-5}$ | 90.4 sec. |

Table 2a    Computer Results of the System Reliability Maximization Problem

| Iteration $k$ | Times of f-value calculated at each iteration | Value of $r_k$ | $R_1$ | $R_2$ | $R_3$ | $R_4$ | $-P$ | $-f$ $(= R_s)$ | Cost |
|---|---|---|---|---|---|---|---|---|---|
| 0 | | $2.214 \times 10^{-2}$ | 0.6 | 0.6 | 0.6 | 0.6 | 0.6647 | 0.8862 | 662.4 |
| 1 | 70 | $2.214 \times 10^{-2}$ | 0.6200 | 0.7150 | 0.5850 | 0.6175 | 0.677501 | 0.924867 | 683.298 |
| 2 | 68 | $5.535 \times 10^{-3}$ | 0.7900 | 0.7900 | 0.6600 | 0.6750 | 0.88815 | 0.970493 | 753.431 |
| 3 | 59 | $1.384 \times 10^{-3}$ | 0.8700 | 0.8799 | 0.7400 | 0.70833 | 0.991246 | 0.991240 | 776.258 |
| 4 | 89 | $3.459 \times 10^{-4}$ | 0.872499 | 0.911250 | 0.791250 | 0.736458 | 0.986439 | 0.996132 | 796.919 |
| 5 | 72 | $8.648 \times 10^{-5}$ | 0.904999 | 0.942749 | 0.767749 | 0.722707 | 0.994712 | 0.997902 | 798.981 |
| 6 | 174 | $2.162 \times 10^{-5}$ | 0.944907 | 0.968158 | 0.730748 | 0.707252 | 0.997960 | 0.999207 | 798.889 |
| 7 | 202 | $5.105 \times 10^{-6}$ | 0.973031 | 0.980765 | 0.715331 | 0.693293 | 0.999216 | 0.999740 | 798.831 |
| 8 | 129 | $1.351 \times 10^{-6}$ | 0.983415 | 0.988173 | 0.710188 | 0.688581 | 0.999691 | 0.999898 | 799.273 |
| 9 | 110 | $3.378 \times 10^{-7}$ | 0.989665 | 0.992636 | 0.704067 | 0.687510 | 0.999878 | 0.999960 | 799.504 |
| 10 | 85 | $8.446 \times 10^{-8}$ | 0.993554 | 0.995303 | 0.703067 | 0.685176 | 0.999952 | 0.999984 | 799.680 |
| 11 | 76 | $2.111 \times 10^{-8}$ | 0.996045 | 0.997208 | 0.702590 | 0.683271 | 0.999981 | 0.999994 | 799.730 |
| 12 | 60 | $5.279 \times 10^{-9}$ | 0.997409 | 0.998117 | 0.702590 | 0.682817 | 0.999992 | 0.999997 | 799.908 |

Table 2b    Computer Results of the System Reliability Maximization Problem

| Iteration k | Times of f-value calculated at each iteration | Value of $r_k$ | $R_1$ | $R_2$ | $R_3$ | $R_4$ | $-P$ | $-f$ $(= R_s)$ | Cost |
|---|---|---|---|---|---|---|---|---|---|
|  |  |  | 0.7 | 0.7 | 0.7 | 0.7 | 0.7161 | 0.9548 | 726.6 |
| 1 | 68 | $1.788 \times 10^{-2}$ | 0.640000 | 0.730000 | 0.610000 | 0.632500 | 0.726307 | 0.936015 | 695.180 |
| 2 | 100 | $4.471 \times 10^{-3}$ | 0.726250 | 0.816250 | 0.711250 | 0.783750 | 0.906109 | 0.992678 | 753.153 |
| 3 | 64 | $1.118 \times 10^{-3}$ | 0.816250 | 0.876250 | 0.791250 | 0.783750 | 0.967770 | 0.992678 | 790.870 |
| 4 | 149 | $2.794 \times 10^{-4}$ | 0.873124 | 0.921249 | 0.766864 | 0.744843 | 0.988329 | 0.996569 | 797.343 |
| 5 | 38 | $6.986 \times 10^{-5}$ | 0.881124 | 0.927249 | 0.763874 | 0.750843 | 0.994753 | 0.996927 | 798.672 |
| 6 | 126 | $1.747 \times 10^{-5}$ | 0.911037 | 0.949323 | 0.740436 | 0.736226 | 0.997489 | 0.998211 | 799.632 |
| 7 | 232 | $4.366 \times 10^{-6}$ | 0.969679 | 0.980016 | 0.692296 | 0.715007 | 0.999295 | 0.999699 | 799.637 |
| 8 | 115 | $1.092 \times 10^{-6}$ | 0.983935 | 0.989126 | 0.687223 | 0.702855 | 0.999731 | 0.999911 | 799.305 |
| 9 | 94 | $2.729 \times 10^{-7}$ | 0.990263 | 0.993277 | 0.685080 | 0.698569 | 0.999894 | 0.999965 | 799.322 |
| 10 | 68 | $6.822 \times 10^{-8}$ | 0.993763 | 0.995777 | 0.684080 | 0.697069 | 0.999958 | 0.999986 | 799.593 |
| 11 | 69 | $1.706 \times 10^{-8}$ | 0.996263 | 0.997206 | 0.682652 | 0.695640 | 0.999983 | 0.999994 | 799.568 |
| 12 | 69 | $4.264 \times 10^{-9}$ | 0.997626 | 0.998399 | 0.682652 | 0.694958 | 0.999993 | 0.999998 | 799.733 |

the number of iterations k increase.  The value of the P-function approaches that of the f-function $(= - R_s)$ as the number of iterations proceed.  Thus the minimization of the P-function will eventually lead to the minimization of f-function.

The values of $r_0$ used in Tables 2a and 2b are determined by

$$f(x) = r_0 \sum_i \frac{1}{g_i(x_0)} \tag{15}$$

where $x_0$ is the initial point.  The basis for this selection procedure is to render the value of the penalty of the constraints to be approximately the same order of magnitude as the value of the f-function at the starting point in the P-function formulation

$$P(x_0, r_0) = f(x_0) + r_0 \sum_i \frac{1}{g_i(x_0)}$$

*Example 6-2.*  In this example the object is to find the optimal component reliabilities, $R_s$, which minimize the cost of the system, i.e., [15]

Minimize

$$C_s = \sum_{i=1}^{4} K_i [\tan (\frac{\pi}{2} R_i)]^{\alpha_i} \tag{16}$$

subject to the constraints

$$R_{s,min} \leq 1 - R_3 [(1 - R_1)(1 - R_4)]^2$$

$$- (1 - R_3)\{1 - R_2[1 - (1 - R_1)(1 - R_4)]\}$$

$$R_{i,min} \leq R_i \leq 1.0, \qquad i = 1,2,3,4$$

The numerical values of parameters are

$$K_1 = 25, \ K_2 = 25, \ K_3 = 50, \ K_4 = 37.5, \ R_{s,min} = 0.99$$

$$R_{i,min} = 0.50, \ \alpha_i = 1.0, \ \text{for } i = 1,2,3,4$$

The cost function suggested in eq. (16) satisfies the following requirements, especially when the reliability of each component, $R_i$, is greater than 0.50.

(1)  Cost of a low reliability component is very low.

(2)  Cost of a high reliability component is very high.

(3)  Cost is a monotonically increasing function of reliability.

(4)  Derivative of cost (with respect to reliability) is a monotonically increasing function of reliability.

The problem is formulated to be solved by SUMT as follows:

Minimize

$$f(x) = C$$

$$= \sum_{i=1}^{4} K_i \left[ \tan \left( \frac{\pi}{2} R_i \right) \right]^{\alpha_i}$$

subject to

$$g_1(x) = 1 - R_3 [(1 - R_1)(1 - R_4)]^2$$

$$- (1 - R_3)\{1 - R_2 [1 - (1 - R_1)(1 - R_4)]\}^2 - R_{s,min} \geq 0$$

$$g_{i+1} = R_i - R_{i,min} \geq 0, \qquad i = 1,2,3,4$$

$$g_{i+5} = 1.0 - R_i \geq 0, \qquad i = 1,2,3,4$$

The P-function for this problem is

$$P(x, r_k) = f(x) + r_k \sum_{i} 1/g_i(x)$$

$$= \sum_{i=1}^{4} K_i \left[ \tan \left( \frac{\pi}{2} R_i \right) \right]^{\alpha_i}$$

$$+ r_k \left\{ \frac{1}{g_1(x)} + \sum_{i=1}^{4} \frac{1}{R_i - R_{i,min}} + \sum_{i=1}^{4} \frac{1}{1.0 - R_i} \right\}$$

where x is the row vector of $(R_1, R_2, R_3, R_4)$.

For this problem, the RAC program fails to satisfy the special requirement that the non-negativity constraints should never be violated during the search.  The results obtained by applying the newly developed program are presented in Tables 3, 4a and 4b.

The solutions obtained from two sets of starting component reliabilities, namely, $[R_1, R_2, R_3, R_4] = [0.6, 0.6, 0.6, 0.6]$ and $[R_1, R_2, R_3, R_4] = [0.5, 0.5, 0.5, 0.5]$, are presented in Table 3. The solutions are almost identical, that is, the optimal minimum cost, C, of 394.806 for the system reliability, $R_s$, of 0.990311 is obtained for the first set of starting components reliabilities, and the optimal minimum cost, C, of 397.879 for the system reliability, $R_s$, Of 0.990406 is obtained for the second set of starting components reliabilities.  Recall that the constraint on the system reliability is 0.99.  The optimal component reliabilities are almost the same for both starting sets.  The stopping criterion for terminating the minimization of the P-function at each iteration is when the number of cut-down step-size operations is 4.  And the final stopping criterion for terminating the problem is $\varepsilon = 10^{-2}$.  For the first set of starting points, it takes 10 iterations for the P-functions, k = 10, with a total of 1135 f-functional values calculated, and for the second set, 11 iterations for the P-functions, k = 11, with a total of 2124 f-functional values calculated.

Results given in Tables 4a and 4b show that the cost of the system, C, is monotonically decreasing as the number of iterations k increase.  The value of the P-function approaches that of the f-function (= C) as the number of iterations proceed.  Thus the minimization of the P-function will eventually lead to the minimization of the f-function.

Again, the values of $r_0$ are determined from eq. (15) as explained in Example  6-1.

It is worth noting that the starting points $R^0 = [R_1, R_2, R_3, R_4] = [0.6, 0.6, 0.6, 0.6]$ and $R^0 = [R_1, R_2, R_3, R_4] = [0.5, 0.5, 0.5, 0.5]$ in Table 4a and Table 4b are in the infeasible region.  The system reliability given by $R^0 = [0.6, 0.6, 0.6, 0.6]$ is 0.8862 and by $R^0 = [0.5, 0.5, 0.5, 0.5]$ is 0.7997, both of which are less than $R_{s,min}$ of 0.99.  Therefore, before the P-function minimization routine

Table 3    Solution of the Cost Minimization Problem (Example 6-2)

| Iteration of $r_k$ k | Number of f-value calculated | Values of component reliabilities | | | | System reliability | Cost | Stopping Criteria | |
|---|---|---|---|---|---|---|---|---|---|
| | | $R_1$ | $R_2$ | $R_3$ | $R_4$ | $R_s$ | | for each k "INCUT" | for final $\varepsilon$ "THETA" |
| 0 | | 0.6 | 0.6 | 0.6 | 0.6 | 0.8862 | 189.3 | | |
| | 1135 | | | | | | | 4 | $10^{-2}$ |
| 10 | | 0.825758 | 0.886685 | 0.645749 | 0.739332 | 0.990311 | 394.806 | | |
| 0 | | 0.5 | 0.5 | 0.5 | 0.5 | 0.7997 | 137.5 | | |
| | 2124 | | | | | | | 4 | $10^{-2}$ |
| 11 | | 0.834409 | 0.879062 | 0.667413 | 0.741163 | 0.990406 | 397.879 | | |

Table 4a    Computer Results of the Cost Function Minimization Problem (Example 6-2)

| Iteration $k$ | Times of f-value calculated at each iteration | Value of $r_k$ | Component reliability | | | | $P$ | f (= cost) | $R_s$ |
|---|---|---|---|---|---|---|---|---|---|
| | | | $R_1$ | $R_2$ | $R_3$ | $R_4$ | | | |
| | | | 0.6 | 0.6 | 0.6 | 0.6 | 596600 | 189.3 | 0.8862 |
| 0 | | | 0.88 | 0.88 | 0.88 | 0.88 | 1737000 | 720.8 | 0.993028 |
| 1 | 128 | 10000 | 0.879375 | 0.932500 | 0.890000 | 0.880000 | 1607460 | 848.322 | 0.999092 |
| 2 | 44 | 2000 | 0.879375 | 0.932500 | 0.890000 | 0.878750 | 322169 | 846.247 | 0.999085 |
| 3 | 60 | 400 | 0.880000 | 0.932500 | 0.87500 | 0.876250 | 65107.1 | 836.419 | 0.999060 |
| 4 | 68 | 80 | 0.878750 | 0.932500 | 0.873759 | 0.871250 | 13673.5 | 796.267 | 0.998937 |
| 5 | 100 | 16 | 0.876250 | 0.930000 | 0.835000 | 0.848750 | 3325.04 | 696.863 | 0.998447 |
| 6 | 110 | 3.3 | 0.864375 | 0.920000 | 0.771875 | 0.809373 | 1156.76 | 568.441 | 0.9970269 |
| 7 | 143 | 0.64 | 0.849062 | 0.907500 | 0.710625 | 0.773125 | 673.137 | 477.356 | 0.994747 |
| 8 | 136 | 0.128 | 0.837187 | 0.898750 | 0.670625 | 0.750626 | 483.416 | 430.118 | 0.992646 |
| 9 | 128 | 0.0256 | 0.827500 | 0.893750 | 0.651875 | 0.743125 | 428.820 | 408.365 | 0.991314 |
| 10 | 218 | 0.00512 | 0.825758 | 0.886685 | 0.645749 | 0.739332 | 411.440 | 394.806 | 0.990311 |

Table 4b    Computer Results of the Cost Function Minimization Problem (Example 6-2)

| Iteration k | Times of f-value calculated at each iteration | Value of $r_k$ | $R_1$ | $R_2$ | $R_3$ | $R_4$ | P | f (=cost) | $R_s$ |
|---|---|---|---|---|---|---|---|---|---|
|  |  |  | 0.5 | 0.5 | 0.5 | 0.5 | $4 \times 10^{53}$ | 137.5 | 0.7997 |
| 0 |  |  | 0.82 | 0.82 | 0.80 | 0.80 | $5.483 \times 10^{7}$ | 441.3 | 0.990738 |
| 1 | 159 | 5000 | 0.878750 | 0.932500 | 0.89000 | 0.880000 | $8.042 \times 10^{5}$ | 847.634 | 0.999089 |
| 2 | 68 | 1000 | 0.879375 | 0.932500 | 0.888750 | 0.878750 | $1.615 \times 10^{5}$ | 842.963 | 0.999078 |
| 3 | 76 | 200 | 0.879375 | 0.932500 | 0.883750 | 0.876250 | $3.297 \times 10^{4}$ | 826.498 | 0.999032 |
| 4 | 107 | 40 | 0.878750 | 0.931250 | 0.862500 | 0.863750 | 7225.49 | 760.683 | 0.998791 |
| 5 | 115 | 8 | 0.871875 | 0.926250 | 0.810000 | 0.833750 | 1994.90 | 640.217 | 0.997972 |
| 6 | 166 | 1.6 | 0.857500 | 0.914375 | 0.743750 | 0.792500 | 849.602 | 522.950 | 0.996097 |
| 7 | 211 | 0.32 | 0.843125 | 0.903750 | 0.688125 | 0.761875 | 550.085 | 452.783 | 0.993768 |
| 8 | 303 | 0.064 | 0.828108 | 0.885363 | 0.673714 | 0.747464 | 467.363 | 406.040 | 0.991086 |
| 9 | 305 | 0.0128 | 0.834409 | 0.879062 | 0.667413 | 0.741163 | 429.843 | 397.879 | 0.990406 |
| 10 | 307 | 0.00256 | 0.834409 | 0.879062 | 0.667413 | 0.741163 | 404.272 | 397.879 | 0.990406 |
| 11 | 307 | 0.000512 | 0.834409 | 0.879062 | 0.667413 | 0.741163 | 399.158 | 397.879 | 0.990406 |

is started, a new feasible point is first searched.  The point [0.88, 0.88, 0.88, 0.88] in the second row of Table 4a is thus selected and is used as the feasible starting point to start the minimization procedure again.  Also, the point [0.82, 0.82, 0.80, 0.80] is selected and used as feasible starting point for Table 4b.

*Example 6-3.*  To demonstrate the technique, the five-stage reliability problem is solved.  The problem is

Maximize

$$R_s = \prod_{j=1}^{5} [1 - (1 - R_j)^{x_j}]$$

subject to

$$g_1 = \sum_{j=1}^{5} p_j (x_j)^2 \leq P$$

$$g_2 = \sum_{j=1}^{5} c_j (x_j + \exp (x_j/4)) \leq C$$

$$g_3 = \sum_{j=1}^{5} w_j x_j \exp (x_j/4) \leq W$$

where $x_j \geq 1$, $j = 1,2,\ldots,5$ are integers.

The constants associated with the five-stage problem are

| $j$ | $R_j$ | $p_j$ | $P$ | $c_j$ | $C$ | $w_j$ | $W$ |
|---|---|---|---|---|---|---|---|
| 1 | 0.80 | 1 | | 7 | | 7 | |
| 2 | 0.85 | 2 | | 7 | | 8 | |
| 3 | 0.90 | 3 | 110 | 5 | 175 | 8 | 200 |
| 4 | 0.65 | 4 | | 9 | | 6 | |
| 5 | 0.75 | 2 | | 4 | | 9 | |

The problem is formulated to solve by SUMT as follows:
Minimize

$$Z = f(x) = - R_s$$

$$= - \prod_{j=1}^{5} [1 - (1 - R_j)^{x_j}]$$

subject to

$$g_1' = P - \sum_{j=1}^{5} P_j(x_j)^2 \geq 0$$

$$g_2' = C - \sum_{j=1}^{5} c_j(x_j + \exp(x_j/4)) \geq 0$$

$$g_3' = W - \sum_{j=1}^{5} w_j x_j \exp(x_j/4) \geq 0$$

$$g_{j+3}' = x_j - 1 \geq 0, \qquad j = 1,2,\ldots,5$$

$$g_9' = Z + 1 \geq 0$$

The P-function for this problem is

$$P(x, r_k) = f(x) + r_k \sum_{i=1}^{9} \frac{1}{g_i(x)}$$

$$= - \prod_{j=1}^{5} [1 - (1 - R_j)^{x_j}] + r_k \left[ \frac{1}{P - \sum_{j=1}^{5} P_j(x_j)^2} \right.$$

$$\left. + \frac{1}{C - \sum_{j=1}^{5} c_j(x_j + \exp(x_j/4))} + \frac{1}{W - \sum_{j=1}^{5} w_j x_j \exp(x_j/4)} \right.$$

$$+ \sum_{j=1}^{5} \frac{1}{x_j - 1} + \frac{1}{Z + 1}]$$

where x is the row vector of $(x_1, x_2, x_3, x_4, x_5)$; each of the components are assumed to be continuous variables.

The solutions obtained from the starting number of components
used at each stage, namely, $(x_1, x_2, x_3, x_4, x_5) = (2, 2, 2, 2, 2)$,
are presented in Table 5.  The stopping criterion for terminating
the minimization of the P-function at each k iteration is when the
number of cut-down step-size operations in the Hooke and Jeeves pattern search is 4, and the final stopping criterion for terminating
the problem is $\varepsilon = 10^{-3}$.  As shown in Table 5, it requires 8 iterations for the P-functions, k = 8, with a total of 1600 f-functional
values evaluated.  The table also shows that the system reliability,
$R_s$, monotonically increases after iteration 2 as iteration k increases.
The value of the P-function approaches that of the f-function $(- R_s)$
as the iteration procedure  proceeds.  Thus the minimization of the
P-function will eventually lead to the minimization of the f-function.

Again, the values of $r_0$ are determined from eq. (15) as explained
in Example 6-1.

The five-stage reliability problem solved by KSU's version of
SUMT gives the optimal system configuration, $(x_1, x_2, x_3, x_4, x_5) =$
(2.691, 2.323, 2.047, 3.521, 2.809).  The system reliability with this
configuration is 0.9229.  However, all $x_j$, j = 1,2,...,5 should be
positive integers, therefore a rounding off procedure to the nearest
integer is required.  Two possible rounding off results may exist,
namely,

(A)  $(x_1, x_2, x_3, x_4, x_5) = (3, 2, 2, 4, 3)$, and

(B)  $(x_1, x_2, x_3, x_4, x_5) = (3, 2, 2, 3, 3)$

Configuration (A) gives a higher system reliability than (B)
(because one more redundancy is used at stage 4); but configuration
(A) consume 111 of $g_1$ which is greater than the available resource,

Table 5    Computer Results of the System Reliability Optimization Problem (Example 6-3)

| Iteration | Times of f-value calculated at each iteration | Value of | | | | | | | | Slacks | | |
|---|---|---|---|---|---|---|---|---|---|---|---|---|
| | | $r_k$ | $x_1$ | $x_2$ | $x_3$ | $x_4$ | $x_5$ | P | $f$ $(=R_s)$ | $g_1$ | $g_2$ | $g_3$ |
| 0 | | 5000 | 2.000 | 2.000 | 2.000 | 2.000 | 2.000 | $4.811 \times 10^4$ | 0.8025 | 52.00 | 35.21 | 37.22 |
| 1 | 267 | 2500 | 2.480 | 3.080 | 3.280 | 1.640 | 2.00 | $2.084 \times 10^4$ | 0.7531 | 33.84 | 37.28 | 10.43 |
| 2 | 185 | 312.5 | 2.563 | 3.070 | 3.230 | 1.615 | 1.855 | $2.602 \times 10^3$ | 0.7397 | 35.97 | 38.05 | 14.05 |
| 3 | 84 | 39.06 | 2.557 | 3.07 | 3.23 | 1.615 | 1.860 | $3.245 \times 10^2$ | 0.7396 | 35.96 | 38.08 | 14.06 |
| 4 | 83 | 4.88 | 2.565 | 3.065 | 3.23 | 1.615 | 1.865 | $3.992 \times 10^1$ | 0.7401 | 35.94 | 38.02 | 13.94 |
| 5 | 184 | 0.610 | 2.610 | 3.045 | 3.200 | 1.635 | 1.895 | 4.339 | 0.7468 | 36.04 | 37.59 | 13.63 |
| 6 | 195 | 0.0763 | 2.748 | 2.910 | 3.025 | 1.795 | 2.150 | 0.1314 | 0.7914 | 35.93 | 35.52 | 12.04 |
| 7 | 302 | 0.00238 | 3.016 | 2.590 | 2.463 | 3.011 | 2.536 | 0.8728 | 0.9078 | 21.15 | 22.63 | 0.39 |
| 8 | 300 | 0.000298 | 2.691 | 2.323 | 2.047 | 3.521 | 2.809 | 0.9175 | 0.9229 | 14.03 | 22.40 | 0.60 |

110. Therefore (A) is not feasible. Under configuration (B), $R_s$ = 0.9045, $g_1$ = 83, $g_2$ = 146.1 and $g_3$ = 194.5. Configuration (B) is then the optimal solution.

## REFERENCES


1. Carroll, C. W., "An operation research approach to the economic optimization of a Kraft Pulping process," Ph.D. Dissertation, Institute of Paper Chemistry, Appletown, Wisc., 1959.

2. Carroll, C. W., "The created response surface technique for optimizing nonlinear restrained systems," *Operations Research*, No. 9, pp 169-184 (1961).

3. Fiacco, A. V., and G. P. McCormick, "The sequential unconstrained minimization technique for nonlinear programming: a primal-dual method," *Management Sci.*, Vol. 10, pp. 360-366 (1964).

4. Fiacco, A. V., and G. P. McCormick, "Computational algorithm for the sequential unconstrainted minimization technique for nonlinear programming," *Management Sci.*, Vol. 10, pp. 601-617 (1964).

5. Fiacco, A. V., and G. P. McCormick, "Extension of SUMT for non-linear programming: equality constraints and extrapolation," *Management Sci.*, Vol. 12, pp. 816-829 (1966).

6. Fiacco, A. V., and G. P. McCormick, *Nonlinear Programming: Sequential Unconstrained Minimization Techniques*, New York: Wiley (1968).

7. Hooke, R., and T. A. Jeeves, "Direct search solution of numerical and statistical problems," *J. Assoc. Compt. Mach.*, Vol. 8, pp. 212 (1961).

8. Hwang, C. L., L. T. Fan, and S. Kumar, "Hooke and Jeeves pattern search solution to optimal production planning problems," Report No. 18, Institute for Systems Design and Optimization, Kansas State University (1969).

9. Hwang, C. L., K. C. Lai, F. A. Tillman, and L. T. Fan, "Optimization of system reliability by the sequential unconstrained minimization technique," *IEEE Trans. on Reliability*, Vol. R-24, pp. 133-135 (1975).

10. Hsu, F. T., L. T. Fan and C. L. Hwang, "Sequential unconstrained minimization technique (SUMT) for optimal production planning," Report No. 26, Institute for Systems Design and Optimization, Kansas State University (1971).

11. Lai, K. C., "Optimization of industrial management systems by the sequential unconstrained minimization technique," M. S. Report, Dept. of Industrial Engineering, Kansas State University (1970).

12. McCormick, G. P., W. C. Mylander, III., and A. V. Fiacco, "RAC computer program implementing the sequential unconstrained minimization technique for nonlinear programming," SHARE Number 3189.

13. Paviani, D. A., and D. M. Himmelblau, "Constrained nonlinear optimization by heuristic programming," *Operations Research,* Vol. 17, pp. 872-882 (1969).

14. Tillman, F. A., C. L. Hwang, L. T. Fan, and K. C. Lai, "Optimal reliability of a complex system," *IEEE Trans. on Reliability,* Vol. R-19, pp. 95-99 (1970).

15. Aggarwal, K. K., and J. S. Gupta, "On minimizing the cost of reliability systems," *IEEE Trans. on Reliability,* Vol. R-24, pp. 205 (1975).

16. Bazovsky, I., *Reliability Theory and Practice,* Englewood Cliffs, N.J.: Prentice-Hall (1961).

## 7.1   INTRODUCTION

The Generalized Reduced Gradient Method (GRG) was proposed by Abadie
and Carpentier [1,2].  The method is a generalization of the Wolfe
reduced gradient method [8,9], which solves problems having a nonlinear
objective function and linear equality constraints. It classifies the
variables either as dependent or independent, and substitutes expressions obtained from the linear equality constraints for the dependent
variables into the objective function in terms of the independent variables.  The original problem is then reduced to one which is unconstrained with smaller dimensions. A variety of optimization techniques
can then be used.  The same concept can be applied to problems with
nonlinear constraints, but the solution may become complicated to the
point that only by using numerical methods can a solution be obtained
[7].

GRG has been studied extensively and coded in FORTRAN by Abadie
[3], Abadie and Guigou [4], and Guigou [5,6].  Three generations of
programs, namely, GRG 66, GRG 69, and GREG, have been developed. The
improved code, GREG, is the outgrowth of the first two codes and pro-
mises to remain among the highly regarded nonlinear programming procedures available.

The algorithm for the generalized reduced gradient method is
presented in Appendix A3.  The content of this chapter is based on
Hwang et al. [11].

## 7.2   NUMERICAL EXAMPLES

*Example 7-1.*  The problem of maximizing the reliability of the complex
system given in Fig. 1 and subject to a single constraint, can be
stated as follows (See Example 6-1 in Chapter 6):
Maximize

$$R_s = 1 - Q_s$$

$$= 1 - R_3 [(1 - R_1)(1 - R_4)]^2$$

$$- (1 - R_3)\{1 - R_2[1 - (1 - R_1)(1 - R_4)]\}^2$$

subject to

$$C_s = \sum_{i=1}^{4} c_i \leq C$$

$$R_{i,min} \leq R_i \leq 1.0$$

where

$$c_i = K_i \left[\tan\left(\frac{\pi}{2} R_i\right)\right]^{\alpha_i}, \qquad i = 1,2,3,4$$

The numerical values of the parameters are

$$k_1 = 25, \ k_2 = 25, \ k_3 = 50, \ k_4 = 37.5, \ C = 800$$

$$R_{i,min} = 0.50, \ \alpha_i = 1.0, \ \text{for } i = 1,2,3,4$$

We can now use the GREG computer code to solve this example.  This
problem is reformulated to be

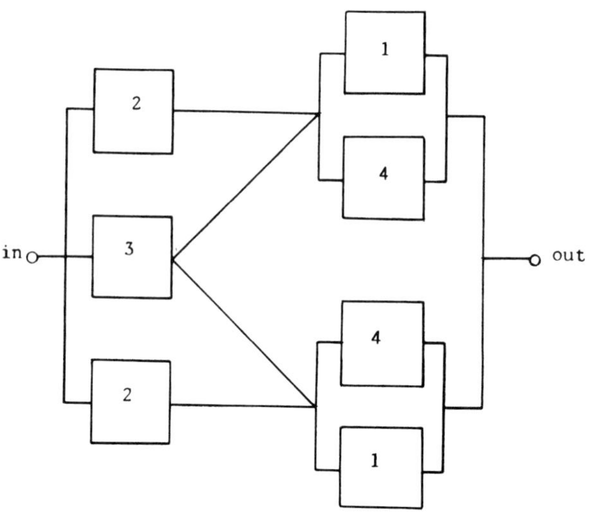

Figure 1    A schematic diagram of a complex system.

Maximize

$$f_0(\bar{x}) = 1 - R_3 \, [(1 - R_1)(1 - R_4)]^2$$

$$- (1 - R_3)\{1 - R_2[1 - (1 - R_1)(1 - R_4)]\}^2$$

subject to

$$f_1(\bar{x}) = \sum_{i=1}^{4} k_i \, [\tan(\tfrac{\pi}{2} R_i)]^{\alpha_i} - 800 \leq 0$$

$$R_{i,min} \leq R_i \leq 1.0, \qquad i = 1,2,3,4$$

In terms of the computer program, four user-supplied subroutines are
used; these are PHIX which defines the objective function, CPHI which
defines the constraint functions, JACOB which defines the gradient
of the constraint functions, and GRADFI which defines the gradient
of the objective function.

Starting with the initial point, $[R_1, R_2, R_3, R_4]^0 = [0.52, 0.52, 0.52, 0.52]$, the solutions are determined to be $[R_1, R_2, R_3, R_4] = [0.902968, 0.948525, 0.813532, 0.851429]$ with the maximum system reliability, $R_s$, of 0.9990396 and has a total cost of 799.949076. By starting with the initial point of $[R_1, R_2, R_3, R_4]^0 = [0.60, 0.60, 0.60, 0.60]$, the solutions are determined to be $[R_1, R_2, R_3, R_4] = [0.896629, 0.949567, 0.830751, 0.842901]$ with the maximum system reliability, $R_s$, of 0.9990471 and a total cost of 799.999023.

*Example 7-2.* The numerical example, Example 7-1, is restated below. The objective is to find the optimal $R_i$'s which minimize the system cost [10]

$$C_s = \sum_{i=1}^{4} k_i \, [\tan (\frac{\pi}{2} R_i)]^{\alpha_i}$$

subject to the constraints

$$R_{s,min} \leq 1 - R_3 \, [(1 - R_1)(1 - R_4)]^2$$

$$- (1 - R_3) \, \{1 - R_2 [1 - (1 - R_1)(1 - R_4)]\}^2$$

$$R_{i,min} \leq R_i \leq 1.0, \qquad i = 1,2,3,4$$

The numerical values of the parameters are

$$k_1 = 25, \ k_2 = 25, \ k_3 = 50, \ k_4 = 37.5$$

$$R_{i,min} = 0.50, \ \alpha_i = 1.0, \ \text{for } i = 1,2,3,4, \ R_{s,min} = 0.99$$

The problem is then reformulated to

Maximize

$$f_0(\bar{x}) = - \sum_{i=1}^{4} k_i [\tan (\frac{\pi}{2} R_i)]^{\alpha_i}$$

subject to

$$f_1(\bar{x}) = -1 + R_3[(1 - R_1)(1 - R_4)]^2$$

$$+ (1 - R_3)\{1 - R_2[1 - (1 - R_1)(1 - R_4)]\}^2 + R_{s,min} \leq 0$$

$$R_{i,min} \leq R_i \leq 1.0, \qquad i = 1,2,3,4$$

Again the four user-supplied subroutines, PHIX, CPHI, JACOB, and GRADFI, are used. Starting with the initial point, $[R_1, R_2, R_3, R_4]^0 = [0.52, 0.52, 0.52, 0.52]$, the solutions are determined to be $[R_1, R_2, R_3, R_4]^* = [0.827672, 0.891787, 0.634234, 0.732349]$ with the minimum cost of 396.85345 and the system reliability of 0.9904930. By starting with the initial point, $[R_1, R_2, R_3, R_4]^0 = [0.60, 0.60, 0.60, 0.60]$, the solutions are determined to be $[R_1, R_2, R_3, R_4] = [0.829047, 0.892711, 0.638432, 0.734509]$ with a minimum cost of 400.79110 and the system reliability of 0.9907858.

*Example 7-3.* To demonstrate the technique of GRG, the previously mentioned five-stage reliability problem is solved which is to Maximize

$$R_s = \prod_{j=1}^{5} [1 - (1 - R_j)^{x_j}]$$

subject to

$$g_1 = \sum_{j=1}^{5} p_j(x_j)^2 \leq P$$

$$g_2 = \sum_{j=1}^{5} c_j(x_j + \exp(x_j/4)) \leq C$$

$$g_3 = \sum_{j=1}^{5} w_j x_j \exp(x_j/4) \leq W$$

where $x_j \geq 1$, $j = 1,2,\ldots,5$ are integers.

The constants associated with the five-stage problem are

| j | $R_j$ | $p_j$ | P | $c_j$ | C | $w_j$ | W |
|---|-------|-------|-----|-------|-----|-------|-----|
| 1 | 0.80 | 1 | | 7 | | 7 | |
| 2 | 0.85 | 2 | | 7 | | 8 | |
| 3 | 0.90 | 3 | 110 | 5 | 175 | 8 | 200 |
| 4 | 0.65 | 4 | | 9 | | 6 | |
| 5 | 0.75 | 2 | | 4 | | 9 | |

Note that in optimizing system reliability, the decision variables, namely, the number of components used at each stage, are considered as continuous variables.  The nearest integer numbers are assigned to them eventually to obtain the required whole component solution.

We now apply GREG to solve this problem.  The example is reformulated to be

Maximize

$$f_0(\bar{x}) = \sum_{j=1}^{5} \ln [1 - (1 - R_j)^{x_j}]$$

subject to

$$f_1(\bar{x}) = \sum_{j=1}^{5} p_j x_j^2 - 110 \leq 0$$

$$f_2(\bar{x}) = \sum_{j=1}^{5} c_j (x_j + \exp (x_j/4)) - 175 \leq 0$$

$$f_3(\bar{x}) = \sum_{j=1}^{5} w_j x_j \exp (x_j/4) - 200 \leq 0$$

Using the four user-supplied subroutines: PHIX, CPHI, JACOB, and GRADFI, we obtain the optimal solution.

The five-stage reliability problem solution has the system configuration of $(x_1, x_2, x_3, x_4, x_5) = (2.678, 2.353, 2.070, 3.531, 2.792)$. The system reliability with this configuration is 0.9235. Since all $x_j$, $j = 1,2,\ldots,5$ should be positive integers, the above results should be rounded off to the nearest integer to be

(A).  $(x_1, x_2, x_3, x_4, x_5) = (3, 2, 2, 4, 3)$, or
(B).  $(x_1, x_2, x_3, x_4, x_5) = (3, 2, 2, 3, 3)$

Configuration (A), although resulting in a higher system reliability, requires 111 of $g_1$ which is greater than the available resource of 110. Configuration (B) yields a system reliability of $R_s = 0.9045$ and requires $g_1 = 83$, $g_2 = 146.1$, and $g_3 = 194.5$. Therefore, configuration (B) is the optimal solution within the constraint set.


## REFERENCES

1.  Abadie, J. and J. Guigou, Numerical experiments with the GRG method, in *Integer and Nonlinear Programming*, J. Abadie (ed.), Amsterdam:  North Holland Publishing Co. (1970).

2.  Abadie, J., and J. Carpenter, Generalization of the Wolfe Reduced Gradient Method to the Case of Nonlinear Constraints, in *Optimization*, R. Fletcher (ed.), New York:  Academic Press (1969).

3.  Abadie, J., Solution des questions de degenerescence dans la methode GRG, EDF* note HI 143/100 due 25 (September 1969).

4.  Abadie, J. and J. Guigou, Gradient refuit generalise, EDF* note HI 069/2 du Avril (1969).

5.  Guigou, J., Presentation et utilisation du code GRG, EDF* note HI 102/02 du 9 Juin (1969).

6.  Guigou, J., Presentation et utilisation du code GREG, EDF* note HI 582/2 due 25 Mai (1971).

7.  Hwang, C. L., J. L. Williams, and L. T. Fan, Introduction to the Generalized Reduced Gradient Method, Institute for Systems Design and Optimization, Report No. 39, Kansas State University, Manhattan, Kansas 66506.

8. Wolfe, P., Methods of nonlinear programming, in: *Recent Advances in Mathematical Programming*, R. Graves and P. Wolfe (eds.), New York: McGraw-Hill (1963).

9. Wolfe, P., Methods for linear constraints, in: *Nonlinear Programming*, J. Abadie (ed.), Amsterdam: North Holland Publishing Co. (1967).

10. Aggarwal, K. K., and J. S. Gupta, "On minimizing the cost of reliable systems," *IEEE Transactions on Reliability*, Vol. R-24, No. 3, p. 205 (1975).

11. Hwang, C. L., F. A. Tillman, and W. Kuo, "Reliability optimization by generalized Lagrangian-function and reduced-gradient methods," *IEEE Transactions on Reliability*, Vol. R-28, No. 4, pp. 316-319 (1979).

CHAPTER 8          METHOD OF LAGRANGE MULTIPLIERS AND THE KUHN-TUCKER
                    CONDITIONS IN OPTIMAL SYSTEMS RELIABILITY

## 8.1  INTRODUCTION

The general nonlinear programming problem can be solved by the method
of Lagrange multipliers when the problem has the following characteristics:  (1)  no inequalities appear in the constraints, (2)  no non-
negativity or discreteness restrictions are imposed on the variables,
(3)  the number of equality constraints is less than the number of
variables, and (4)  the objective and constraint functions are continuous and possess partial second order derivates. The necessary and
sufficient conditions are developed from Taylor's series expansion [6].

The method of Lagrange multipliers can be generalized to solve
problems involving inequality constraints and non-negative variables.
The necessary conditions for optimizing the problems are the classical
Kuhn-Tucker conditions.  The necessary conditions are also sufficient
for a global minimum, if the objective function is a convex function
and the constraints form a convex set in a feasible region, and for
a global maximum if the objective function is a concave function and
the constraints form a convex set in a feasible region.

Wolfe based on the Kuhn-Tucker conditions, introduced the modified
simplex method for quadratic programming problems. This is widely used
and simple to apply.  Many authors have derived the necessary and
sufficient conditions for various cases of nonlinear programming problems utilizing the Kuhn-Tucker conditions.  For details, see [6].

Several papers have addressed the application of the method of
Lagrange multipliers in conjunction with the Kuhn-Tucker conditions

for the following reliability optimization problems:

*Example 8-1.* *(A Single Constraint Problem):* Given a system reliability requirement $R_{s,min}$, the problem is to determine a least-cost allocation of an N-stage series system that yields $R_s \geq R_{s,min}$. As an example, consider the following four-stage system with a system reliability requirement of $R_{s,min} = 0.99$ and total cost less than $b_1 = 61$:

| Stage, j | 4 | 3 | 2 | 1 |
|---|---|---|---|---|
| $c_j$ | 1.2 | 2.3 | 3.4 | 4.5 |
| $R.$ | 0.8 | 0.7 | 0.75 | 0.85 |

The problem is to

Maximize

$$R_s = \prod_{j=1}^{N} [1 - (1 - R_j)^{x_j}]$$

subject to

$$g_1 = \sum_{j=1}^{N} c_j x_j \leq b_1$$

and

$$R_s \geq R_{s,min}$$

*Example 8-2.* *(Two Linear Constraint Problem):* Consider an example of a series system of four stages. The component reliability, cost, and weight data are:

| Stage, j | 1 | 2 | 3 | 4 |
|---|---|---|---|---|
| Component reliability, $R_j$ | 0.80 | 0.70 | 0.75 | 0.85 |
| Cost, $c_j$ | 1.2 | 2.3 | 3.4 | 4.5 |
| Weight, $w_j$ | 5 | 4 | 8 | 7 |

The system cost and weight are 56 and 120, respectively.

The problem is to

Maximize

$$R_s = \prod_{j=1}^{4} [1 - (1 - R_j)^{x_j}]$$

subject to

$$g_1 = \sum_{j=1}^{4} c_j x_j \leq 56$$

$$g_2 = \sum_{j=1}^{4} w_j x_j \leq 120$$

where $x_j \geq 1$, $j = 1,2,3,4$ are integers.

A simple Lagrange multiplier method may be used to solve a single constraint problem, e.g., Example 8-1. In this approach, trial and error is used until all resources are consumed, where the number of redundancies is assumed to be continuous even though it must be an integer. However, it is difficult to use Lagrange multipliers with multiple constraints, e.g., Example 8-2. To solve Example 8-2, the Kuhn-Tucker conditions are used to generate a set of simultaneous equations which can be solved by Newton's method. The solution yields an unique value of the Lagrange multipliers. Theoretically, a nonlinear constraint problem can also be solved by the Lagrange multiplier method and the Kuhn-Tucker conditions.

## 8.2  THE METHOD OF LAGRANGE MULTIPLIERS FOR A SINGLE CONSTRAINT PROBLEM

Again Example 8-1 in the Introduction section is considered. For this single constraint problem, one Lagrange multiplier, $\lambda$, should be introduced to form the unconstrained maximum of the new function

$$L(\bar{x}) = R_s - \lambda \left( \sum_{j=1}^{N} c_j x_j - b_1 \right) \tag{1}$$

A solution which maximizes the Lagrangian of eq. (1) is the solution
to the constrained maximization problem. The constraint is the amount
of the resource expended. In general, difference choices of the $\lambda$'s
lead to different resource levels, and it may be necessary to adjust
them by trial and error to utilize the maximum allowable resource, $b_1$.
Therefore, an adjustment of the $\lambda$'s is required [4].

Since the maximization of the logarithm of the system reliability
maximizes the objective function, we transform the problem to be the
logarithm of the reliability.

$$H = \ln R_s$$

$$= \sum_{j=1}^{N} \ln [1 - (1 - R_j)^{x_j}] \tag{2}$$

For a given $\lambda$, the Lagrange multiplier function will be formed as

$$L(\bar{x}) = \sum_{j=1}^{N} \ln [1 - (1 - R_j)^{x_j}] - \lambda \left( \sum_{j=1}^{N} c_j x_j - b_1 \right) \tag{3}$$

over the integers $x_j \geq 1$, $j = 1, 2, \ldots, N$.

Eq. (3) can be maximized in the usual fashion by differentiation with
respect to $x_j$ and equating to zero to obtain the optimal $x_j$, then
rounding off the values to the nearest integers. Namely,

$$\frac{dL(\bar{x})}{dx_j} = 0 \tag{4}$$

or

$$\frac{-(1 - R_j)^{x_j} \ln (1 - R_j)}{[1 - (1 - R_j)^{x_j}]} - \lambda c_j = 0 \tag{5}$$

which leads to the (real) solution for $x_j$:

$$x_j = \frac{\ln \{1/[1 - \ln (1 - R_j)/\lambda c_j]\}}{\ln (1 - R_j)} \tag{6}$$

for each stage.  The rounding off procedures to $x_j$, $j = 1,2,3,4$ to upper and lower nearest integers are tested to determine which results in the greatest value of $R_s$, within the constraint limitations.

Referring to Example 8-1, application of the Lagrange multiplier method as previously developed for a series of values of $\lambda$ produces the solutions shown in Table 1.  Inspection of the results shows that in all but one case the changes in allocation from one solution to the next consists of at most one additional component in at most one stage.  Therefore the reliability and cost are monotonically increasing with the number of components used and there is no $\lambda$ which could produce new solutions between these solutions.  However, the transition from $\lambda = 0.0003$ to $0.0002$ produced a change in three stages, and we can expect further solutions in this interval for intermediate $\lambda$ values.  Additional exploration of this region yields two more solutions, as given in Table 2.

Since there are no longer any changes by more than one component between successive solutions, the optimal allocation is $[x_1, x_2, x_3, x_4]^* = [5, 7, 6, 4]$ with the system reliability, $R_s = 1 - 0.001288 = 0.998712$, and the cost $\sum\limits_{j=1}^{4} c_j w_j = 60.5$, which is the same result obtained by using the dynamic programming approach.


## 8.3   THE KUHN-TUCKER CONDITIONS

The Kuhn-Tucker conditions can be stated as follows [6,7]:
A point $(x_1, x_2, \ldots, x_n)$ which optimizes a function

$$S = f(x_1, x_2, \ldots, x_n) \tag{7}$$

subject to the inequality constraints

$$g_j(x_1, x_2, \ldots, x_n) \leq 0, \qquad j = 1,2,\ldots,r \tag{8}$$

Table 1

| $\lambda$ | COST | SYSTEM UNRELIABILITY | ALLOCATION | | | |
|---|---|---|---|---|---|---|
| | | | Stage 1 | Stage 2 | Stage 3 | Stage 4 |
| 0.0009 | 44.6 | 0.009997 | 5 | 5 | 4 | 3 |
| 0.0008 | 48.0 | 0.007086 | 5 | 5 | 5 | 3 |
| 0.0007 | 50.3 | 0.005392 | 5 | 6 | 5 | 3 |
| 0.0006 | 54.8 | 0.002530 | 5 | 6 | 5 | 4 |
| 0.0005 | 54.8 | 0.002530 | 5 | 6 | 5 | 4 |
| 0.0004 | 54.8 | 0.002530 | 5 | 6 | 5 | 4 |
| 0.0003 | 54.8 | 0.002530 | 5 | 6 | 5 | 4 |
| 0.0002 | 61.7 | 0.001033 | 6 | 7 | 6 | 4 |

Table 2

| $\lambda$ | COST | SYSTEM UNRELIABILITY | ALLOCATION | | | |
|---|---|---|---|---|---|---|
| | | | Stage 1 | Stage 2 | Stage 3 | Stage 4 |
| 0.000225 | 54.8 | 0.002530 | 5 | 6 | 5 | 4 |
| 0.000220 | 57.1 | 0.002020 | 5 | 7 | 5 | 4 |
| 0.000215 | 60.5 | 0.001288 | 5 | 7 | 6 | 4 |
| 0.000210 | 61.7 | 0.001033 | 6 | 7 | 6 | 4 |

exists if there is a set of $\lambda_1$, $\lambda_2$,..., $\lambda_m$ that satisfies the following set of conditions.

$$\frac{\partial L}{\partial x_i} = \frac{\partial f}{\partial x_i} - \sum_{j=1}^{r} \lambda_j \frac{\partial g_j}{\partial x_i} = 0, \qquad i = 1,2,\ldots,n \qquad (9)$$

$$\lambda_j g_j = 0, \qquad j = 1,2,\ldots,r \qquad (10)$$

$$g_j \leq 0, \qquad j = 1,2,\ldots,r \qquad (11)$$

$$\lambda_j \geq 0, \qquad j = 1,2,\ldots,r \ \text{(for maximization)} \qquad (12a)$$

or

$$\lambda_j \leq 0, \qquad j = 1,2,\ldots,r \ \text{(for minimization)} \qquad (12b)$$

These conditions are also sufficient for a global minimum if f and $g_j$, $j = 1,2,\ldots,r$ are all convex and differentiable functions and for a global maximum if f is concave and $g_j$, $j = 1,2,\ldots,r$ are all convex and differentiable functions.

Similarly, the necessary conditions for optimization of the function given by eq. (7), subject to the inequality constraints, eq. (8), and the constraint of non-negative x are

$$\frac{\partial L}{\partial x_i} = \frac{\partial f}{\partial x_i} - \sum_{j=1}^{r} \lambda_j \frac{\partial g_j}{\partial x_i} \leq 0, \qquad i = 1,2,\ldots,n$$
$$\text{(for maximization)} \qquad (13a)$$

or

$$\frac{\partial L}{\partial x_i} = \frac{\partial f}{\partial x_i} - \sum_{j=1}^{r} \lambda_j \frac{\partial g_j}{\partial x_i} \geq 0, \qquad i = 1,2,\ldots,n$$
$$\text{(for minimization)} \qquad (13b)$$

$$x_i \frac{\partial L}{\partial x_i} = 0, \qquad i = 1,2,\ldots,n \qquad (14)$$

$$\lambda_j g_j = 0, \qquad j = 1,2,\ldots,r \qquad (15)$$

$$g_j \leq 0, \qquad j = 1,2,\ldots,r \qquad (16)$$

$$x_i \geq 0, \qquad i = 1,2,\ldots,n \qquad\qquad (17)$$

$$\lambda_j \geq 0, \qquad j = 1,2,\ldots,r \text{ (for maximization)} \qquad (18a)$$

or

$$\lambda_j \leq 0, \qquad j = 1,2,\ldots,r \text{ (for minimization)} \qquad (18b)$$

Eqs. (12a), (12b), (18a) and (18b) are based on the fact that
if $\lambda > 0$, the stationary point cannot be a minimum, and if $\lambda < 0$, it
can not be a maximum [7]. Note that the sign of $\lambda$ will be affected
by factors such as the nature of the optimization problem (whether
maximization or minimization), the type of inequality constraints
[whether $g_j(x) \leq 0$ or $g_j(x) \geq 0$], and the form of the Lagrangian func-
tion [whether $L(x, \lambda) = f(x) - \sum_j \lambda_j g_j(x)$ or $L(x,\lambda) = f(x) + \sum_j \lambda_j g_j(x)$].
Recall that eqs. (12a), (12b), (18a) and (18b) are based on the in-
equality constraints given by eq. (8) [$g_j(x) \leq 0$], and the Lagrangian
function of the form, $L(x,\lambda) = f(x) - \sum_j \lambda_j g_j(x)$.

## 8.4 METHOD OF LAGRANGE MULTIPLIERS AND THE KUHN-TUCKER CONDITIONS FOR THE TWO LINEAR CONSTRAINT PROBLEM

When more than one constraint is imposed on the problem, the
trial and error procedure for searching for the $\lambda$'s associated with
each constraint is not practical. In this example, the Kuhn-Tucker
conditions are applied to simplify the problem [9,10].

For an N-stage series system, the problem can be restated as

Maximize

$$R_s = \prod_{j=1}^{N} [1 - (1 - R_j)^{x_j}] \qquad\qquad (19)$$

subject to

$$\sum_{j=1}^{N} a_{ij} x_j \leq b_i, \qquad i = 1,2,\ldots,r \qquad\qquad (20)$$

If we denote $(1 - R_j)$ by $Q_j$, $(1 - R_j)^{x_j}$ by $Q'_j$, then eq. (19) becomes

$$R_s = \sum_{j=1}^{N} (1 - Q'_j)$$

Since maximization of the logarithm of the system reliability maximizes the objective function [1], we can denote the objective function by

$$\ln R_s = \sum_{j=1}^{N} \ln (1 - Q'_j) \tag{21}$$

Also, by setting

$$Q_j^{x_j} = Q'_j$$

we obtain

$$x_j = \frac{\ln Q'_j}{\ln Q_j} \tag{22}$$

Substituting $x_j$ into eq. (20) yields

$$\sum_{j=1}^{N} a_{ij} \frac{\ln Q'_j}{\ln Q_j} \le b_i, \qquad i = 1,2,\ldots,r$$

or

$$\sum_{j=1}^{N} \alpha_{ij} \ln Q'_j \le b_i, \qquad i = 1,2,\ldots,r \tag{23}$$

where

$$\alpha_{ij} = \frac{a_{ij}}{\ln Q_j} \tag{24}$$

Since the objective (eq. (21)) and the constraint functions (eq. (23)) are all separable concave and convex of $Q'_j$ respectively, this guarantees the global maximum [5].

The Lagrange function, whose stationary point is to be found, is

$$L(R', \lambda) = \sum_{j=1}^{N} \ln R'_j - \sum_{i=1}^{r} \lambda_i \left[ \sum_{j=1}^{N} (\alpha_{ij} \ln (1 - R'_j)) - b_i \right] \quad (25)$$

where $R'_j = 1 - Q'_j$, $Q'_j = (1 - R_j)^{x_j}$.

The Kuhn-Tucker conditions can be written as

$$\frac{\partial L}{\partial R'_j} = \frac{1}{R'_j} + \sum_{i=1}^{r} \lambda_i \alpha_{ij} / (1 - R'_j) = 0, \qquad j = 1,2,\ldots,N \quad (26)$$

$$\lambda_i \left[ \sum_{j=1}^{N} (\alpha_{ij} \ln (1 - R'_j)) - b_i \right] = 0, \qquad i = 1,2,\ldots,r \quad (27)$$

$$\sum_{j=1}^{N} \alpha_{ij} \ln (1 - R'_j) - b_i \leq 0 \quad (28)$$

$$\lambda_i \geq 0, \qquad i = 1,2,\ldots,r \quad (29)$$

Eqs. (26) through (29) form the basis for the optimal solution. A set of N + r equations represented by eqs. (26) and (27) can be solved by Newton's method. Actually, eq. (29) will form the stopping criterion for the iteration process.

After the solution is obtained from the simultaneous equations formed from the Kuhn-Tucker conditions, we combine the solution with eqs. (21) and (22) to obtain the optimal system reliability and the optimal number of redundancies for each stage in which the rounding off procedure, to the nearest integer number, is required.

Referring to Example 8-2, the problem can be stated as
Maximize

$$\ln R_s = \sum_{j=1}^{4} \ln R'_j$$

subject to

$$\sum_{j=1}^{4} \frac{c_j}{\ln(1 - R_j)} \ln(1 - R_j') - C \leq 0$$

$$\sum_{j=1}^{4} \frac{w_j}{\ln(1 - R_j)} \ln(1 - R_j') - W \leq 0$$

The Lagrange function is

$$L(R', \lambda) = \sum_{j=1}^{4} \ln R_j' - \lambda_1 \left[ \sum_{j=1}^{4} \frac{c_j}{\ln(1 - R_j)} \ln(1 - R_j') - C \right]$$

$$- \lambda_2 \left[ \sum_{j=1}^{4} \frac{w_j}{\ln(1 - R_j)} \ln(1 - R_j') - W \right]$$

The Kuhn-Tucker conditions are

$$\frac{\partial L}{\partial R_j'} = \frac{1}{R_j'} + \frac{\lambda_1 c_j + \lambda_2 w_j}{(1 - R_j') \ln(1 - R_j)} = 0, \qquad j = 1,2,3,4 \qquad (30)$$

$$\lambda_1 \left[ \sum_{j=1}^{4} \frac{c_j}{\ln(1 - R_j)} \ln(1 - R_j') - C \right] = 0 \qquad (31)$$

$$\lambda_2 \left[ \sum_{j=1}^{4} \frac{w_j}{\ln(1 - R_j)} \ln(1 - R_j') - W \right] = 0 \qquad (32)$$

$$\sum_{j=1}^{4} \frac{c_j}{\ln(1 - R_j)} \ln(1 - R_j') - C \leq 0 \qquad (33)$$

$$\sum_{j=1}^{4} \frac{w_j}{\ln(1 - R_j)} \ln(1 - R_j') - W \leq 0 \qquad (34)$$

$$\lambda_1, \; \lambda_2 \geq 0 \qquad (35)$$

This problem is solved by the Lagrange multipliers method and the Kuhn-Tucker conditions following eqs. (30) through (35) with the results, $[R_1, R_2, R_3, R_4]^* = [0.999735, 0.999494, 0.999294, 0.999388]$, $[\lambda_1, \lambda_2]^* = [0.00019994, -0.00003730]$, and the system reliability $R_s$, 0.997914. By using eq. (22), we find that the optimal allocation is $[x_1, x_2, x_3, x_4]^* = [5.11, 6.30, 5.23, 3.90]$, which has to be rounded off to the nearest integers e.g. [5, 6, 5, 4]. The system reliability for the allocation of [5, 6, 5, 4] is 0.997471, and requires $g_1 = 54.8$ and $g_2 = 117$ [10].

## 8.5   CONCLUSION

The Kuhn-Tucker conditions do provide assist in the determination of the optimal solution.  However, it is not always possible to solve large scale nonlinear programming problems with this approach.  Another caution is that it is not necessarily true that every point which is a solution to the Kuhn-Tucker conditions will be a point at which the objective function takes on its relative maximum or minimum for all x which satisfy the constraints.  But note that every point at which the objective function assumes its relative maximum or minimum for x satisfying the constraints must also be a solution to the Kuhn-Tucker conditions.  There are many other valuable applications of the Kuhn-Tucker conditions.  An example is in quadratic programming.

## REFERENCES


1.   R. E. Barlow, F. Proschan, and L. C. Hunter, *Mathematical Theory of Reliability,* New York:  Wiley (1965).

2.   Black, G., and F. Proschan, "On Optimal redundancy," *Operations Research,* Vol. 7, pp. 581-588 (1959).

3.   Bodin, L. D., "Optimization procedure for the analysis of coherent structures," *IEEE Transactions on Reliability,* Vol. R-18, No. 3, pp. 118-126 (1969).

4.  Everett, H. III, "Generalized Lagrange Multiplier Method for
    solving problems of optimum allocation of resources," *Operations
    Research,* Vol. 11, No. 3, pp. 399-417 (1963).

5.  Hadley, G., *Nonlinear and Dynamic Programming,* Reading, Mass.:
    Addison-Wesley (1964).

6.  Hwang, C. L., P. K. Gupta, and L. T. Fan, Method of Lagrange
    Multipliers and the Kuhn-Tucker Conditions, Institute for Sys-
    tems Design and Optimization, Report No. 60, Kansas State Uni-
    versity (1974).

7.  Kuhn, N. W. and A. W. Tucker, Nonlinear Programming, in *Proceed-
    ings of the Second Berkeley Symposium on Mathematical Statistics
    and Probability,* J. Neyman (ed.), Berkeley, California: Univer-
    sity of California Press, pp. 481-492 (1951).

8.  Messinger, M., and H. Shooman, "Technique for optimum spares
    allocation:  a tutorial review," *IEEE Transactions on Reliability,*
    Vol. R-19, pp. 156-166 (1970).

9.  Misra, K. B., and M. D. Ljubojevic, "Optimal reliability design
    of a system: a new look," *IEEE Transactions on Reliability,* Vol.
    R-22, pp. 255-258 (1973).

10. Misra, K. B., "Reliability optimization of a series-parallel sys-
    tem," *IEEE Transactions on Reliability,* Vol. R-21, No. 4, pp.
    230-238 (1972).

11. Shershin, A. C., "Mathematical optimization techniques for the
    simultaneous apportionments of reliability and maintainability,"
    *Operations Research,* Vol. 18, No. 1, pp. 95-106 (1970).

12. Tucker, A. W., "Linear and nonlinear programming," *Operations
    Research,* Vol. 5, pp. 244-257 (1957).

CHAPTER 9          THE GENERALIZED LAGRANGIAN FUNCTION METHOD APPLIED
                   TO OPTIMAL SYSTEMS RELIABILITY


## 9.1  INTRODUCTION

A general class of mathematical programming problems can be stated
as

    Problem (A):   minimize $f(\bar{x})$                              (1)

subject to

$$g_i(\bar{x}) \geq 0, \qquad i = 1,\ldots,m \tag{2}$$

$$\bar{x} \,\varepsilon\, \Omega \tag{3}$$

where $\bar{x} \,\varepsilon\, E^n$, and $\Omega$ is a subset of n-Euclidean space $E^n$.  In addition
it is assumed that $f(\bar{x})$, and $g_1(\bar{x})$, $g_2(\bar{x})$, $\ldots$, $g_m(\bar{x})$ are real value
functions on $\Omega$ and are twice continuously differentiable.

Problem (A) can be solved by methods which are based on the transformation of the constrained problem into a sequence of unconstrained
problems.  There are two classes of methods, namely, the penalty and
the Lagrangian methods.  The penalty methods (e.g., sequential unconstrained minimization technique, see Chapter 6), have been studied extensively and used to solve many practical problems [3,5].  However,
they suffer from numerical instability.  The Lagrange multiplier methods have been used mostly for the analysis of economic systems [2].
Recently, augmented Lagrangian functions have been proposed to solve
problems with equality [6,7,10] and inequality constraints [1,11,12,
13,14].

In this Chapter a new version of the generalized or augmented
Lagrangian function proposed by Sayama et al.  [12,13] is used for
finding the solution of a nonlinear programming problem with inequality constraints and is applied to solving optimal system reliability
problems.  The function is twice continuously differentiable and clo-
sely related to the generalized penalty function which includes the
interior and exterior penalty functions as special cases.  The content
of this chapter is based on Hwang et al. [15].

The theoretical properties of the function and the computational
algorithm are presented in [13].  The method has been shown to be locally convergent to the saddle points of the generalized Lagrangian.
By using this method, we can find the Lagrange multipliers associated
with the solution of problem (A).  Thus we can solve problems which
play an important part in design and synthesis in the fields of engineering and economics.


## 9.2   THE GENERALIZED LAGRANGIAN FUNCTION AND ITS COMPUTATIONAL PROCEDURES

The classical Lagrangian function associated with Problem (A)
is defined as

$$L(\bar{x}, \bar{\lambda}) = f(\bar{x}) - \sum_{i=1}^{m} \lambda_i g_i(\bar{x}) \tag{4}$$

where $\lambda_i$, $i = 1,2,\ldots,m$ are the Lagrange multipliers.  The literature
on the penalty method and the method of Lagrange multipliers is well
reviewed in Fiacco and McCormick [5], Lootsma [8], and Rockafellar
[11].

Although several examples have been suggested to satisfy the
properties of the generalized Lagrangian, a proper choice of the function is of utmost importance in obtaining an efficient method of
solution.  A class of the generalized or augmented Lagrangian proposed

by Sayama et al. [12,13] is

$$
L(\bar{x}, \bar{\lambda}; t) = f(\bar{x}) - \sum_{i=1}^{m}
\begin{cases}
\lambda_i g_i(\bar{x}) - t g_i^2(\bar{x}), & g_i(\bar{x}) \leq 0 \\[2ex]
\dfrac{(\lambda_i)^2 g_i(\bar{x})}{\lambda_i + t g_i(\bar{x})}, & g_i(\bar{x}) \geq 0
\end{cases}
\tag{5}
$$

or in a similar form to the classical Lagrangian

$$
L(\bar{x}, \bar{\lambda}; t) = f(\bar{x}) - \sum_{i=1}^{m} \lambda_i g_i(\bar{x}) + \sum_{i=1}^{m}
\begin{cases}
t g_i^2(\bar{x}), & g_i(\bar{x}) \leq 0 \\[2ex]
\dfrac{\lambda_i t g_i^2(\bar{x})}{\lambda_i + t g_i(\bar{x})}, & g_i(\bar{x}) \geq 0
\end{cases}
\tag{6}
$$

where $\lambda_i$, $i = 1,2,\ldots,m$ are multipliers and $t > 0$ is a penalty parameter. $L(\bar{x}, \bar{\lambda}; t)$ is termed the multiplier function, and the computational algorithm using the function is called the multiplier method.
$L(\bar{x}, \bar{\lambda}; t)$ is constructed in such a way that it is twice continuously
differentiable if $f(\bar{x})$, and $g_i(\bar{x})$, $i = 1,2,\ldots,m$ are twice continuously
differentiable. This property is quite important to the computational
procedure for finding the unconstrained minimum of the generalized
Lagrangian.

It is worth noting that by letting $t = 0$ in $L(\bar{x}, \bar{\lambda}; t)$, eq. (5)
is reduced to the classical Lagrangian, eq. (4). The multiplier function can also be interpreted as an exterior penalty function if $\lambda_i = 0$,
$i = 1,2,\ldots, m$ in $L(\bar{x}, \bar{\lambda}; t)$.

A computational algorithm which makes use of the multiplier function associated with Problem (A) is considered. The penalty parameter, $t$, if chosen sufficiently large (say $10^5$), is kept constant.
Let $\bar{\lambda}^1$ be an initial estimate of $\bar{\lambda}$, and let $\bar{x}^k$ denote a point which
minimizes $L(\bar{x}, \bar{\lambda}^k; t)$; i.e.,

$$L_x(\bar{x}^k, \bar{\lambda}^k; t) = \nabla f(\bar{x}^k) - \sum_{i=1}^{m} \left\{ \begin{array}{l} [\lambda_i^k - 2tg_i(\bar{x}^k)]\nabla g_i(\bar{x}^k) \\[2em] \dfrac{(\lambda_i^k)^3}{[\lambda_i^k + tg_i(\bar{x}^k)]^2}\,\nabla g_i(\bar{x}^k) \end{array} \right\} = 0 \qquad (7)$$

This suggests that we take

$$\lambda_i^{k+1} = \left\{ \begin{array}{ll} \lambda_i^k - 2tg_i(\bar{x}^k), & g_i(\bar{x}^k) \leq 0, \\[2em] \dfrac{(\lambda_i^k)^3}{[\lambda_i^k + tg_i(\bar{x}^k)]^2}, & g_i(\bar{x}^k) \geq 0, \end{array} \right. \qquad i = 1,\ldots,m, \qquad (8)$$

so that $(\bar{x}^k, \bar{\lambda}^{k+1})$ satisfies the following equation:

$$L_x(\bar{x}^k, \bar{\lambda}^{k+1}) = \nabla f(\bar{x}^k) - \sum_{i=1}^{m} \lambda_i^{k+1} \nabla g_i(\bar{x}^k) = 0 \qquad (9)$$

If $\bar{\lambda}^1 \geq 0$, then $\bar{\lambda}^k$ is forced to be non-negative according to the correction of eq. (8). Eq. (8) may be represented as follows:

$$\lambda_i^{k+1} = \left\{ \begin{array}{ll} \lambda_i^k - 2\xi_i tg_i(\bar{x}^k), & g_i(\bar{x}^k) \leq 0 \\[2em] \lambda_i^k - \xi_i \dfrac{t\lambda_i^k g_i(\bar{x}^k)\,[2\lambda_i^k + tg_i(\bar{x}^k)]}{[\lambda_i^k + tg_i(\bar{x}^k)]^2}, & g_i(\bar{x}^k) \geq 0 \end{array} \right. \qquad i = 1,\ldots,m \qquad (10)$$

where $1 \geq \xi_i > 0$. If $\xi_i = 1$, eq. (10) is equivalent to eq. (8). By using the multiplier function, eq. (10) can be written as follows:

$$\lambda_i^{k+1} = \lambda_i^k + cL_{\lambda_i}(\bar{x}^k, \bar{\lambda}^k; t), \qquad i = 1,\ldots,m \qquad (11)$$

where c is a scalar,

$$
c = \begin{cases}
2\xi_i t \\[2em]
2\xi_i t \; \dfrac{\lambda_i^k + 0.5tg_i(\bar{x}^k)}{\lambda_i^k + 2tg_i(\bar{x}^k)}
\end{cases}
$$

The computational procedure by the multiplier method may be summarized as follows:

1) Choose a penalty parameter $t > 0$ and initial values of the multiplier $\bar{\lambda}^1 > 0$.

2) Find the $\bar{x}^k$ that minimizes $L(\bar{x}, \bar{\lambda}^k; t)$. Any multidimensional search technique, e.g., the sequential simplex pattern search, may be used.

3) Stop the iterations when one of the following criteria is satisfied.

$$
|\lambda_i^k g_i(\bar{x}^k)| \le \varepsilon, \qquad i = 1,\ldots,m
$$

or

$$
|f(\bar{x}^k) - L(\bar{x}^k, \bar{\lambda}^k; t)| \le \varepsilon
$$

where $\varepsilon$ is a sufficiently small positive number.

4) Select $\bar{\lambda}^{k+1}$ by eq. (8) or eq. (10), and return to Step 2).

## 9.3   NUMERICAL EXAMPLES

*Example 9-1.*   To demonstrate the generalized Lagrangian function method, the five-stage reliability problem is solved. The problem is to

Maximize

$$R_s = \prod_{j=1}^{5} [1 - (1 - R_j)^{x_j}]$$

subject to

$$g_1 = \sum_{j=1}^{5} p_j (x_j)^2 \le P$$

$$g_2 = \sum_{j=1}^{5} c_j (x_j + \exp(x_j/4)) \le C$$

$$g_3 = \sum_{j=1}^{5} w_j x_j \exp(x_j/4) \le W$$

where $x_j \ge 1$, $j = 1,2,\ldots,5$ are integers.

The constants associated with the five-stage problem are

| $j$ | $R_j$ | $P_j$ | $P$ | $c_j$ | $C$ | $w_j$ | $W$ |
|---|---|---|---|---|---|---|---|
| 1 | 0.80 | 1 | | 7 | | 7 | |
| 2 | 0.85 | 2 | | 7 | | 8 | |
| 3 | 0.90 | 3 | 110 | 5 | 175 | 8 | 200 |
| 4 | 0.65 | 4 | | 9 | | 6 | |
| 5 | 0.75 | 2 | | 4 | | 9 | |

It is noted that in optimizing the system reliability, the decision
variables, namely, the number of components used at each stage, are
considered to be continuous variables.  The nearest integer solutions
are selected eventually.

To solve this problem, we first must reformulate it as follows:
Minimize

$$f_o(\bar{x}) = - R_s = - \sum_{j=1}^{5} [1 - (1 - R_j)^{x_j}]$$

subject to

$$g_1(\bar{x}) = P - \sum_{j=1}^{5} p_j(x_j)^2 \geq 0$$

$$g_2(\bar{x}) = C - \sum_{j=1}^{5} c_j(x_j + \exp(x_j/4)) \geq 0$$

$$g_3(\bar{x}) = W - \sum_{j=1}^{5} w_j x_j \exp(x_j/4) \geq 0$$

$$g_{3+j}(x_j) = x_j - 1.0 \geq 0, \qquad j = 1,2,\ldots,5$$

$$g_9(\bar{x}) = 1.0 + f_o(\bar{x}) \geq 0$$

The constant penalty parameter, $t = 1.0 \times 10^5$, and the initial estimate of the multipliers $\bar{\lambda}^1 = (1.0, 1.0, 1.0, 1.0, 1.0, 1.0, 1.0, 1.0, 1.0)^T$ are chosen [Step 1]. The generalized Lagrangian function is given by

$$L(\bar{x}, \bar{\lambda}; t) = f_o(\bar{x}) - \sum_{i=1}^{9} \begin{cases} 1 * g_i(\bar{x}) - 1.0 \times 10^5 * g_i^2(\bar{x}), & g_i(x) \leq 0 \\[2em] \dfrac{(1)^2 * g_i(\bar{x})}{1 + 1.0 \times 10^5 * g_i(\bar{x})}, & g_i(\bar{x}) > 0 \end{cases} \tag{12}$$

which is a function of $\bar{x} = (x_1, x_2, x_3, x_4, x_5)^T$ only. The sequential simplex pattern search method starting from $\bar{x}^0 = (1.0, 1.0, 1.0, 1.0, 1.0)^T$ is used to find the minimum $L(\bar{x}, \bar{\lambda}; t)$ of eq. (12) [Step 2]. $\bar{x}^1 = (x_1^1, x_2^1, x_3^1, x_4^1, x_5^1)^T$ is found to be (2.0594, 2.5178, 2.7202, 3.4229, 2.6118) which gives

$$L^1(\bar{x}^1, \bar{\lambda}; t) = -0.9014323$$

$$f_0(\bar{x}^1) = -0.9013424$$

$$g_1(\bar{x}^1) = 13.39341$$

$$g_2(\bar{x}^1) = 26.07391$$

$$g_3(\bar{x}^1) = 15.60743$$

$$g_4(\bar{x}^1) = 1.0594$$

$$g_5(\bar{x}^1) = 1.5178$$

$$g_6(\bar{x}^1) = 1.7202$$

$$g_7(\bar{x}^1) = 1.4299$$

$$g_8(\bar{x}^1) = 1.6118$$

$$g_9(\bar{x}^1) = 0.0986577$$

The stopping criteria are $\varepsilon_1 = 1.0 \times 10^{-3}$ and $\varepsilon_2 = 1.0 \times 10^{-4}$. Since

$$|\lambda_i^1 g_i(\bar{x}^1)| = |1 * g_i(\bar{x}^1)| \leq 1.0 \times 10^{-3}, \text{ for } i = 1,2,\ldots,9$$

and

$$\left| \frac{L(\bar{x}^1, \bar{\lambda}^1; t) - L(\bar{x}^0, \bar{\lambda}^0; t)}{L(\bar{x}^1, \bar{\lambda}^1; t)} \right| = \left| \frac{0.9013426 - 0.9}{0.9013426} \right|$$

$$= 0.0015 \leq 1 \times 10^{-4} \qquad [\text{Step 3}].$$

We should go to Step 4 and choose new $\bar{\lambda}^2$'s using eq. (8). Then go to Step 2.

The iterative procedure is carried out until the stopping criteria are satisfied [Step 3]. The results are presented in Table 1.

The optimal results are $\bar{x} = (x_1, x_2, x_3, x_4, x_5)^T = (2.408,$ 2.376, 2.019, 3.632, 2.898), $f_0(\bar{x}) = -0.9193794$, $g_1 = 12.85325$, $g_2 = 23.77317$, $g_3 = 0.18543$.

Since the number of components used at each stage should be positive integers, the optimal values of $\bar{x}$ shall be rounded off to the nearest integers, and the constraints checked. The possible system configurations are:

(A)  $(x_1, x_2, x_3, x_4, x_5) = (2, 2, 2, 4, 3)$

(B)  $(x_1, x_2, x_3, x_4, x_5) = (3, 2, 2, 4, 3)$

(C)  $(x_1, x_2, x_3, x_4, x_5) = (3, 2, 2, 3, 3)$

Configuration (A) results in a system reliability of $R_s = 0.9007759$, $g_1(\bar{x}) = 4.0$, $g_2(\bar{x}) = 24.74182$ and $g_3(\bar{x}) = 1.7710$. Configuration (B) results in a system reliability of $R_s = 0.9308028$, $g_1(\bar{x}) = -1.0$, $g_2(\bar{x}) = 14.467$, and $g_3(\bar{x}) = -19.6139$ in which constraints 1 and 3 are violated. Configuration (C) results in a system reliability of $R_s = 0.9044667$, $g_1(\bar{x}) = 27$, $g_2(\bar{x}) = 28.879$, and $g_3(\bar{x}) = 7.519$. With configuration (B), the cost exceed the total available. With configuration (A), the system reliability is less than that from configuration (C); therefore configuration (C) is the optimal solution.

*Example 9-2.* The objective of this example is to find the optimal $R_i$'s which minimize

$$C = 2K_1 R_1^{\alpha_1} + 2K_2 R_2^{\alpha_2} + K_3 R_3^{\alpha_3} + 2K_4 R_4^{\alpha_4}$$

subject to the constraints

$$R_{s,min} \leq 1 - R_3 [(1 - R_1)(1 - R_4)]^2$$
$$- (1 - R_3) \{1 - R_2 [1 - (1 - R_1)(1 - R_4)]\}^2$$

$$R_i \geq R_{i,min}, \qquad i = 1,2,3,4$$

Table 1    Computational Results of the System Reliability Maximization Problem (Example 9-1).

$\varepsilon_1 = 10^{-3}$, $\varepsilon_2 = 10^{-4}$, $t = 10^5$, $\lambda_i^1 = 1.0$, and $x_i^o = 1.0$, for $i = 1,2,\ldots,5$

Iteration

| $k$ | $x_1$ | $x_2$ | $x_3$ | $x_4$ | $x_5$ | $-L$ | $R_s=-f_o$ | $g_1^k$ | $g_2^k$ | $g_3^k$ |
|---|---|---|---|---|---|---|---|---|---|---|
| 0 | 1.0 | 1.0 | 1.0 | 1.0 | 1.0 | 0.9 | 0.29835 | 98.0 | 101.91 | 151.21 |
| 1 | 2.0594 | 2.5178 | 2.7207 | 3.4299 | 2.6188 | 0.9013426 | 0.9013423 | 13.39 | 26.07 | 1.32 |
| 2 | 2.2845 | 2.4699 | 2.2402 | 3.3416 | 3.0026 | 0.9166631 | 0.9166632 | 14.83 | 25.08 | 0.20 |
| 3 | 2.3672 | 2.4509 | 1.9734 | 3.7306 | 2.7458 | 0.9182468 | 0.9182469 | 9.95 | 22.27 | 2.04 |
| 4 | 2.3808 | 2.3486 | 2.0810 | 3.6134 | 2.8736 | 0.9182124 | 0.9182125 | 11.57 | 23.36 | 1.49 |
| 5 | 2.3928 | 2.3606 | 2.0930 | 3.5964 | 2.9090 | 0.9192970 | 0.9192970 | 11.33 | 23.06 | 0.0016 |
| 6 | 2.4079 | 2.3757 | 2.0187 | 3.6316 | 2.8983 | 0.9193794 | 0.9193794 | 11.13 | 22.83 | 0.20 |

Computation Time = 5.36 sec.

The numerical values of the parameters are

$$K_1 = 100, \quad K_2 = 100, \quad K_3 = 200, \quad K_4 = 150, \quad \alpha_i = 0.6, \quad i = 1,2,3,4$$

$$R_{s,min} = 0.9, \quad R_{i,min} = 0.5, \quad i = 1,2,3,4$$

The problem is formulated in the generalized Lagrangian function format as follows:

Minimize

$$f_o(\bar{R}) = C$$

$$= 2K_1 R_1^{\alpha_1} + 2K_2 R_2^{\alpha_2} + K_3 R_3^{\alpha_3} + 2K_4 R_4^{\alpha_4}$$

subject to the constraints

$$g_1(\bar{R}) = 1 - R_3 [(1 - R_1)(1 - R_4)]^2$$

$$- (1 - R_3)\{1 - R_2 [1 - (1 - R_1)(1 - R_4)]\}^2 - R_{s,min} \geq 0$$

$$g_{i+1}(\bar{R}) = R_i - R_{i,min} \geq 0, \qquad i = 1,2,3,4$$

$$g_{i+5}(\bar{R}) = 1 - R_i \geq 0, \qquad i = 1,2,3,4$$

$$g_{10}(\bar{R}) = R_3 \ [(1 - R_1)(1 - R_4)]^2$$

$$+ (1 - R_3)\{1- R_2 [1 - (1 - R_1)(1 - R_4)]\}^2 \geq 0$$

where $1 - g_{10}(\bar{R})$ is the system reliability of the complex configuration shown in Fig. 1, and $\bar{R} = [R_1, R_2, R_3, R_4]^T$. $R_1$, $R_2$, $R_3$, and $R_4$ are the reliabilities of components 1, 2, 3, and 4, respectively.

For this example, the SUMT-RAC program failed to satisfy the special requirement that the violable non-negativity constraints should never be violated during the search. The results from applying the generalized Lagrangian function method are presented in Table 2.

Table 2　Computer Results of the Cost Function Minimization Problem (Example 9-2)

$\epsilon_1 = 10^{-3}$, $\epsilon_2 = 10^{-4}$, $t = 10^5$, $\lambda_i = 1.0$, and $R_i = 0.7$ for $i = 1,2,3,4$

Iteration

| k | $R_1$ | $R_2$ | $R_3$ | $R_4$ | $R_s$ | L | $f_o$ |
|---|---|---|---|---|---|---|---|
| 0 | 0.7 | 0.7 | 0.7 | 0.7 | 0.8274 | 700. | 726.61 |
| 1 | 0.5 | 0.83076 | 0.49992 | 0.50021 | 0.8979 | 641.32 | 640.8625 |
| 2 | 0.50004 | 0.83811 | 0.5 | 0.50001 | 0.8999 | 641.82 | 641.7678 |
| 3 | 0.50001 | 0.84062 | 0.5 | 0.5 | 0.9005 | 641.85 | 642.0446 |

Computation Time = 7.65 sec.

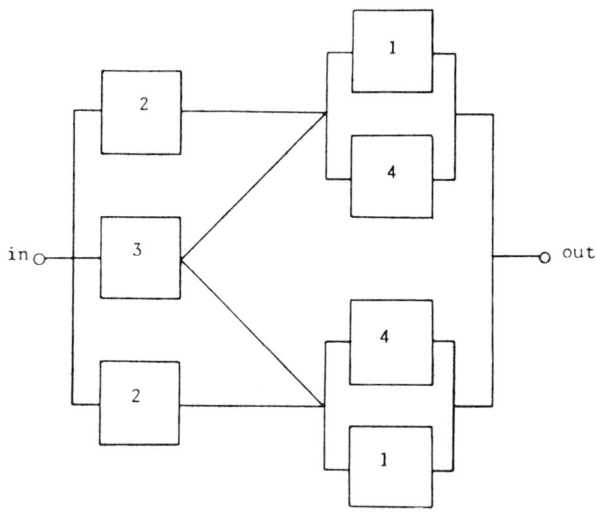

Figure 1    A schematic diagram of a complex system.

The optimal solution obtained from the starting point of $(R_1,$ $R_2,$ $R_3,$ $R_4)^0$ = (0.7, 0.7, 0.7, 0.7) is $(R_1,$ $R_2,$ $R_3,$ $R_4)*$ = (0.50001, 0.84062, 0.5, 0.5), with an optimal minimum cost of C = 642.0446 and the system reliability of $R_s$ = 0.9005.  The penalty parameter t = 1.0 x $10^5$, and the stopping criteria $\varepsilon_1$ = 1.0 x $10^{-3}$ and $\varepsilon_2$ = 1.0 x $10^{-4}$ are used.

It is noted that when the penalty parameter, say t = 1.0 x $10^5$, is large enough, we will finally reach the optimal solution in the feasible region.  Comparing the results obtained by SUMT-KSU (See Chapter 6), the cost is almost the same (C = 642.0446 by this method, and 642.428 by SUMT-KSU), but the system reliability, $R_s$, is slightly higher, $R_s$ = 0.9005 by this method and $R_s$ = 0.900021 by SUMT-KSU's method.  The multiplier method also exhibits a much faster convergence (11.53 sec for this problem) than SUMT-KSU (about 45 sec for the same problem) using an IBM 370/158 computer.

## REFERENCES


1.  Arrow, K. J., F. J. Gould and S. M. Howe, "A general saddle point
    result for constrained optimization," Institute of Statistics
    Mimeo Series No. 774, Univ. of N. Carolina (Chapel Hill) (1971).

2.  Arrow, K. J., L. Hurwicz and H. Uzawa, *Studies in Linear and
    Nonlinear Programming,* Stanford California: Stanford Univ. Press
    (1958).

3.  Bracken, J., and G. P. McCormick, *Selected Applications of Non-
    linear Programming,* New York: Wiley (1968).

4.  Everett, H., "Generalized Lagrange multiplier method for solving
    problems of optimal allocation of resources," *Operations Research,*
    Vol. 11, pp. 399-417 (1963).

5.  Fiacco, A. V., and G. P. McCormick, *Nonlinear Programming, Se-
    quential Unconstrained Minimization Techniques,* New York: Wiley
    (1968).

6.  Haarhoff, P. C., and J. D. Buys, "A new method for the optimization
    of a nonlinear function subject to nonlinear constraints," *Comput.
    J.,* Vol. 13, pp. 178-184 (1970).

7.  Hestenes, M. R., "Multiplier and gradient method," *J. Optim.
    Theory and Applns.,* Vol. 4, pp. 303-320 (1969).

8.  Lootsma, F. A., "A survey of methods for solving constrained
    minimization problems via unconstrained minimizations," in
    *Numerical Methods for Nonlinear Optimization,* F. A. Lootsma (ed.),
    New York:  Academic Press (1973).

9.  Polyak, B. T., "Iteration methods using Lagrange multipliers for
    solving extremal problems with constraints of the equation type,"
    *USSR Comput. Math. and Math. Phys.,* Vol. 10-5, pp. 42-52, (1970).

10. Powell, M. J. D., "A method for nonlinear constraints in minimi-
    zation problems," in *Optimization,* R. Fletcher (ed.), New York:
    Academic Press (1969).

11. Rockafellar, R. T., "Penalty method and augmented Lagrangians in
    nonlinear programming," Proceedings of 5th IFIP Conference on
    Optimization Techniques, Rome, (May 1973).

12. Sayama, H., Y. Kameyana, H. Nakayama, and Y. Sawaragi, "Iteration
    process in Lagrange multipliers for multiplier method," *System
    and Control (Japan),* Vol. 17, pp. 776-778 (1973).

13. Sayama, H., Y. Kameyama, H. Nakayama, and Y. Sawaragi, "The generalized Lagrangian functions for mathematical programming problems," Report 55, Institute for Systems Design and Optimization, Kansas State University (1974).

14. Pierre, D. A., and M. J. Lowe, *Mathematical Programming via Augmented Lagrangians - An Introduction with Computer Programs,* Reading, Massachusetts:  Addison-Wesley (1975).

15. Hwang, C. L., F. A. Tillman, and W. Kuo, "Reliability optimization by generalized Lagrangian-function and reduced-gradient methods," *IEEE Transactions on Reliability,* Vol. R-28, pp. 316-319 (1979).

## 10.1    INTRODUCTION

By using the well-known fact that the arithmetic mean is at least as
great as the geometric mean, dual problems for a variety of primal
design problems may be formulated.  Geometric programming exploits
this implied inequality and the relationship between the primal and
dual problem to facilitate the solution of an optimization problem.
The principal problem must be expressed in terms of a class of functions
which are called positive polynomials or posynomials for short.

In the primal problem, the posynomial is minimized subject to
constraints of the posynomial type.  Because of the inequality relating
the arithmetic and geometric means, there exists a related problem
which requires maximization of the so-called dual function subject
to certain linear constraints [1,2,6].

Geometric programming differs from other optimization techniques
in that it gives the minimum values of $S(x)$ for the posynomial $S$ (primal function) without first locating the point $x$ where $S$ is minimum.
The dual problem is solved first; then the optimal solution of the
primal problem is obtained by the corresponding relationship.  This
is explained in the following sections.

## 10.2   FORMULATION OF THE PROBLEM

A general primal minimization problem involving posynomials subject to r inequality constraints and the corresponding dual problem of maximizing the dual function subject to its constraints can be stated as follows:

*Primal Problem:*
Minimize

$$S = \sum_{j=1}^{n_0} u_j \tag{1}$$

subject to

$$g_i \leq 1, \qquad i = 1,2,\ldots,r \tag{2}$$

where

$$g_i = \sum_{j=m_i}^{n_i} u_j \tag{3}$$

Here

$$m_i = n_{i-1} + 1, \qquad i = 1,2,\ldots,r \tag{4}$$

and the $u_j$ are numbered consecutively from 1 to $n_r = n$.  The $u_j$ are defined:

$$u_j = c_j \prod_{i=1}^{m} x_i^{a_{ji}}, \qquad j = 1,2,\ldots,n \tag{5}$$

where

$$x_1, x_2, \ldots, x_m > 0 \tag{6}$$

The components $a_{ji}$ are real numbers, but the coefficients $c_j$ are assumed to be positive.

The posynomial S which is to be minimized is a function of m independent variables $x_1$, $x_2$, ...,$x_m$. The inequality constraints, eq. (2), are called forced constraints, where the inequality constraints given in eq. (6) are considered to be natural constraints. The matrix $(a_{ji})$ is called the exponent matrix and has n rows and m columns.

The dual problem which corresponds to the primal problem is stated as follows:

*Dual Problem:*

Maximize

$$v = \left[ \prod_{j=1}^{n} \left( \frac{c_j}{\delta_j} \right)^{\delta_j} \right] \prod_{i=1}^{r} \lambda_i^{\lambda_i} \tag{7}$$

where

$$\lambda_i = \sum_{j=m_i}^{n_i} \delta_j, \qquad i = 1,2,\ldots,r \tag{8}$$

Here

$$m_1 = n_0 + 1, \; m_2 = n_1 + 1, \; \ldots, \; m_r = n_{r-1} + 1$$

The constants $c_j$ are assumed to be positive and the weights $\delta_1$, $\delta_2$, ..., $\delta_n$ are subject to the linear constraints:

$$\delta_1 \geq 0, \; \delta_2 \geq 0, \; \ldots, \; \delta_n \geq 0 \tag{9}$$

$$\sum_{j=1}^{n_0} \delta_j = 1 \tag{10}$$

$$\sum_{j=1}^{n} a_{jk} \delta_j = 0, \qquad k = 1,2,\ldots,m \tag{11}$$

where the coefficients, $a_{jk}$, are all real numbers.

The dual function, $v$, is a function of the variables, $\delta_1$, $\delta_2$, ..., $\delta_n$. The linear constraints of the positivity condition, eq. (9), the normality condition, eq. (10), and the orthogonality condition, eq. (11) are imposed on these variables.

Note the manner in which the dual problem is generated from its corresponding primal problem. The positive constants, $c_j$, appearing in the dual function, $v$, are the coefficients of the posynomials whose terms are given by eq. (5). Each $\delta_j$ is associated with the $j^{th}$ terms, $u_j$, of the primal problem, and hence, each $u_j$ of the posynomials is associated with one and only one of the dual variables, $\delta_1$, $\delta_2$, ..., $\delta_n$. Each $\lambda_i$ in the dual problem comes from a forced constraint, $g_i \le 1$, of the primal problem. Because the normality condition forces the weights of the objective function to sum to unity, the $\lambda_0$ corresponding to the objective function itself is unity, and thus does not appear in eq. (7). This normality condition is the only part of the dual problem that distinguishes between the objective function, $S$, and a set of the inequality constraints, $g_i < 1$. The coefficient matrix $(a_{jk})$ that appears in the orthogonality condition, eq. (11), is the exponent matrix of the primal problem.

Since the optimal redundancy allocation problems under consideration have positive coefficients with the variables in the objective function and if the objective function can be transformed to a polynomial form and all the resource requirements associated with each component of the $i^{th}$ constraint and the $j^{th}$ resource have positive values, the geometric programming problem of the "minimization of a posynomial subject to inequality constraints" is formulated and solved [3,4].

Referring to an N-stage parallel redundant system which has linear and separable constraints, the system reliability can be stated as

Maximize

$$R_s = \sum_{j=1}^{N} (1 - Q_j^{x_j}) \tag{12}$$

subject to

$$\prod_{j=1}^{N} p_{ij} x_j \le b_i, \qquad i = 1, 2, \ldots, r \tag{13}$$

when $Q_j \le 0.5$, which is a reasonable assumption for the component un-reliability, then eq. (12) can be approximated as

Maximize

$$R_s \approx 1 - \sum_{j=1}^{N} Q_j^{x_j}$$

or, equivalently,

Minimize

$$S = \sum_{j=1}^{N} Q_j^{x_j} \tag{14}$$

Since the stage unreliability is defined as

$$Q_j' = Q_j^{x_j} \tag{15}$$

then eq. (14) can be expressed as

$$S = \sum_{j=1}^{N} Q_j' \tag{16}$$

Also, from the definition of eq. (15),

$$\ln Q_j' = x_j \ln Q_j$$

or

$$x_j = \frac{\ln Q_j'}{\ln Q_j}, \qquad j = 1, 2, \ldots, N \tag{17}$$

Substituting $x_j$ into eq. (13), we obtain

$$\sum_{j=1}^{N} p_{ij} \frac{\ln Q_j'}{\ln Q_j} \leq b_i, \qquad i = 1,2,\ldots,r$$

Dividing both sides by $b_i > 0$, we obtain

$$\sum_{j=1}^{N} \frac{p_{ij}}{b_i \ln Q_j} \ln Q_j' \leq 1$$

or

$$\sum_{j=1}^{N} \left(- \frac{p_{ij}}{b_i \ln Q_j}\right) (-1) \ln Q_j' \leq 1 \tag{18}$$

If we define

$$k_{ij} = - \frac{p_{ij}}{b_i \ln Q_j} \tag{19}$$

then

$$\sum_{j=1}^{N} (-k_{ij}) \ln Q_j' \leq 1$$

or

$$\sum_{j=1}^{N} \ln Q_j'^{-k_{ij}} \leq 1 \tag{20}$$

or

$$\ln \prod_{j=1}^{N} Q_j'^{-k_{ij}} \leq 1 \tag{21}$$

Taking exponentials of both sides, we have

$$\prod_{j=1}^{N} Q_j^{-k_{ij}} \leq e$$

or

$$e^{-1} \prod_{j=1}^{N} Q_j'^{-k_{ij}} \leq 1, \qquad i = 1,2,\ldots,r \tag{22}$$

The primal geometric programming problem is therefore formulated where we minimize eq. (16) subject to eq. (22).

Assuming $x_j$ to be continuous variables, the dual geometric programming formulation is

Maximize

$$v = \prod_{j=1}^{N} \left[\frac{1}{\delta_j}\right]^{\delta_j} \prod_{i=N+1}^{N+r} \left[\frac{e^{-1}}{\delta_i}\right]^{\delta_i} \prod_{i=1}^{r} (\lambda_i)^{\lambda_i} \tag{23}$$

subject to

$$\sum_{j=1}^{N} \delta_j = 1 \tag{24}$$

and

$$\delta_j - \sum_{i=N+1}^{N+r} k_{ij}\delta_i = 0, \qquad j = 1,2,\ldots,N \tag{25}$$

$$\lambda_i = \sum_{j=m_i}^{n_i} \delta_j, \qquad i = 1,2,\ldots,r \tag{26}$$

$$\delta_i \geq 0, \qquad i = 1,2,\ldots,r \tag{27}$$

where $\delta_i$, $i = 1,2,\ldots,N$ are the dual variables corresponding to eq. (16) and $\delta_{N+j}$, $j = 1,2,\ldots,r$ are the dual variables corresponding to eq. (22).

Eqs. (24) and (25) can be simultaneously solved to obtain $\delta_j^*$, $j = 1,2,\ldots,N+r$. Substitution of these results into eq. (23) gives rise to $v(\delta^*)$ where $v(\delta^*) = v(\delta_1^*, \delta_2^*, \ldots, \delta_{N+r}^*)$. It has been proven [1] that

$$S(Q^*) = v(\delta^*) \tag{28}$$

and

$$u_j(Q'^*) = \delta_j^* S(Q'^*), \qquad j = 1,2,\ldots,N \tag{29}$$

where

$$u_j = Q_j', \qquad j = 1,2,\ldots,N \tag{30}$$

From eqs. (28) and (28), the optimal values of $Q_j'$, $j = 1,2,\ldots,N$ can be obtained.

Finally, using eq. (15) we find the optimal allocation, $x_j$, $j = 1, 2,\ldots,N$.


## 10.3   A NUMERICAL EXAMPLE

Consider the problem in which $N$ stages are connected in series and redundant components, $x_j - 1$, are added in parallel at each stage. The objective is to determine $x_j$ at each stage, such that the system reliability is maximized and the cost and the weight constraints are not exceeded.  The problem is to

Maximize

$$R_s = \prod_{j=1}^{N} [1 - (1 - R_j)^{x_j}] \tag{31}$$

subject to

$$\left. \begin{aligned} g_1 &= \sum_{j=1}^{N} c_j x_j \leq C \\[2ex] g_2 &= \sum_{j=1}^{N} w_j x_j \leq W \end{aligned} \right\} \tag{32}$$

The constants associated with this problem are given as

| Stage | Cost | | Weight | | Probability of Survival |
| --- | --- | --- | --- | --- | --- |
| $j$ | $c_j$ | $C$ | $w_j$ | $W$ | $R_j$ |
| 1 | 1.2 | | 1.0 | | 0.80 |
| 2 | 2.3 | | 1.0 | | 0.70 |
| 3 | 3.4 | 56.0 | 1.0 | 30.0 | 0.75 |
| 4 | 4.5 | | 1.0 | | 0.85 |

The objective function can also be stated as

Minimize

$$Z = \sum_{j=1}^{4} Q'_j \tag{33}$$

subject to

$$e^{-1} \prod_{j=1}^{4} Q'_j{}^{-k_{ij}} \leq 1, \qquad i = 1,2 \tag{34}$$

where

$$Q'_j = Q_j{}^{x_j} = (1 - R_j)^{x_j} \tag{35}$$

$$k_{1j} = -\frac{c_j}{C \ln Q_j}, \qquad j = 1,2,3,4 \tag{36}$$

$$k_{2j} = -\frac{w_j}{W \ln Q_j}, \qquad j = 1,2,3,4 \tag{37}$$

The dual function is

$$v = \left(\frac{1}{\delta_1}\right)^{\delta_1} \left(\frac{1}{\delta_2}\right)^{\delta_2} \left(\frac{1}{\delta_3}\right)^{\delta_3} \left(\frac{1}{\delta_4}\right)^{\delta_4} \left(\frac{e^{-1}}{\delta_5}\right)^{\delta_5} \left(\frac{e^{-1}}{\delta_6}\right)^{\delta_6} (\lambda_1)^{\lambda_1} (\lambda_2)^{\lambda_2} \tag{38}$$

where

$$\lambda_1 = \delta_5$$

$$\lambda_2 = \delta_6$$

The normality condition becomes

$$\delta_1 + \delta_2 + \delta_3 + \delta_4 = 1 \tag{39}$$

and the orthogonality conditions become

$$\left.\begin{aligned}
\delta_1 - [k_{11}\delta_5 + k_{21}\delta_6] &= 0 \\[2mm]
\delta_2 - [k_{12}\delta_5 + k_{22}\delta_6] &= 0 \\[2mm]
\delta_3 - [k_{13}\delta_5 + k_{23}\delta_6] &= 0 \\[2mm]
\delta_4 - [k_{14}\delta_5 + k_{24}\delta_6] &= 0
\end{aligned}\right\} \tag{40}$$

Various methods can be applied to find the optimal $\delta_1^*$, $\delta_2^*$, $\delta_3^*$, $\delta_4^*$, $\delta_5^*$, $\delta_6^*$, and $v^*$.

From eqs. (39) and (40) we can express $\delta_1$, $\delta_2$, $\delta_3$, $\delta_4$, and $\delta_5$ in terms of $\delta_6$:

$$\delta_j = \frac{k_{1j}}{\sum\limits_{j=1}^{4} k_{1j}} + \left(k_{2j} - \frac{k_{1j}\sum\limits_{j=1}^{4} k_{2j}}{\sum\limits_{j=1}^{4} k_{1j}}\right)\delta_6, \qquad j = 1,2,3,4 \tag{41}$$

$$\delta_5 = \frac{1}{\sum\limits_{j=1}^{4} k_{1j}} - \frac{\sum\limits_{j=1}^{4} k_{2j}}{\sum\limits_{j=1}^{4} k_{1j}}\delta_6 \tag{42}$$

Substituting eqs. (41) and (42) into eq. (38), the objective function is then a one-dimensional function, in term of $\delta_6$, to be maximized. The Golden Section method can be applied to find

$$v^* = 0.00207$$

$$\delta_6^* = 0.0046$$

By substituting into eqs. (41) and (42), we can obtain,

$$\delta_1^* = 0.09969$$

$$\delta_2^* = 0.25537$$

$$\delta_3^* = 0.32786$$

$$\delta_4^* = 0.31707$$

$$\delta_5^* = 7.48310$$

Therefore,

$$Q_1' = 0.2^{x_1} = \delta_1^* v^* = 0.0002064$$

$$Q_2' = 0.3^{x_2} = \delta_2^* v^* = 0.0005286$$

$$Q_3' = 0.25^{x_3} = \delta_3^* v^* = 0.0006787$$

$$Q_4' = 0.15^{x_4} = \delta_4^* v^* = 0.0006563$$

From these equations, the optimal $x_j$, $j = 1,2,3,4$ are found to be

$$x_1 = 5.27248$$

$$x_2 = 6.26699$$

$$x_3 = 5.26248$$

$$x_4 = 3.86317$$

which gives the system reliability, $R_s = 0.99793$.  After rounding off to the nearest integer number, we find the optimal allocation:

$$x_1 = 5$$

$$x_2 = 6$$

$$x_3 = 5$$

$$x_4 = 4$$

The system reliability is 0.99747 with slack in the cost constraint of 1.4 and in the weight constraint of 10.

## REFERENCES


1.  Duffin, R. J., E. L. Peterson, and C. Zener, *Geometric Programming,* New York: Wiley (1967).

2.  Fan, L. T., L. E. Erickson, and C. L. Hwang, Geometric Programming, Institute for Systems Design and Optimization, Vol. 5, Kansas State University, Manhattan, Kansas  66502 (1973).

3.  Federowicz, A. J., and M. Mazumdar, "Use of geometric programming to maximize reliability achieved by redundancy," *Operations Research,* Vol. 16, No. 5, pp. 948-954 (1968).

4.  Misra, K. B., and J. Sharma, "A new geometric programming formulation for a reliability problem," *International Journal of Quality Control,* Vol. 18, No. 3, pp. 497-503 (1973).

5.  Nijkamp, P., *Planning of Industrial Complexes by Means of Geometric Programming,* Rotterdam, Netherlands:  Rotterdam University Press (1972).

6.  Zener, C., *Engineering Design by Geometric Programming,* New York: Wiley-Interscience (1971).

## 11.1  INTRODUCTION

In many problems the decision variables must have integer values to
make any sense.  The redundancy allocation problem for system reliability is a good example.

To use some solution techniques the decision variables are assumed
to be continuous, even though they must be in fact integer numbers.
The solution is obtained by rounding the fractional values of the optimal solution to integer values.  However, this approach has its risks.

Various papers have presented the application of integer programming to a variety of problems.  Problems treated in these papers can
be classified into the following types:

*Example 11.1.*    (Linear Objective Function):  The problem is to minimize a linear cost function

$$f = \sum_{j=1}^{N} c_j m_j$$

of an N-stage series system, where $m_j+1$ components are used in the
$j^{th}$ stage subject to the constraints:

$$R_s \geq R_{s,min}$$

$$\sum_{j=1}^{N} w_j m_j \leq W$$

where

$$R_s = \prod_{j=1}^{N} [1 - (1 - R_j)^{m_j+1}]$$

The constants associated with the problem are given as

$$N = 2, \ R_{s,min} = 0.9903, \ W = 40, \ R_1 = 0.91, \ R_2 = 0.96$$

$$c_1 = 5, \ c_2 = 8, \ w_1 = 9, \ w_2 = 6$$

*Example 11.2.* (Nonlinear Objective Function and Linear Constraint Functions): Consider the problem in which N stages are connected in series and redundant components, $m_j$, are added in parallel at each stage. The objective is to determine $m_j$ at each stage, such that the system reliability is maximized and the weight and cost constraints are not exceeded. The problem is stated as

Maximize

$$R_s = \prod_{j=1}^{N} [1 - (1 - R_j)^{m_j+1}]$$

subject to

$$g_1 = \sum_{j=1}^{N} c_j m_j \le C$$

$$g_2 = \sum_{j=1}^{N} w_j m_j \le W$$

Two sets of data listed below, have previously been associated with this problem. The cost, weight, and probability of·survival are given for the redundancies at each stage.

A.   Consider the data originally presented in [25, 26]:

| Stage | Cost | | Weight | | Probability of Survival |
|---|---|---|---|---|---|
| $j$ | $c_j$ | C | $w_j$ | W | $R_j$ |
| 1 | 5 | | 8 | | 0.90 |
| 2 | 4 | | 9 | | 0.75 |
| 3 | 9 | 100 | 6 | 104 | 0.65 |
| 4 | 7 | | 7 | | 0.80 |
| 5 | 7 | | 8 | | 0.85 |

The problem is also restricted so that

$$0 \le m_j \le 4, \qquad j = 1,2,\ldots,5$$

B.   Consider the data used in [4, 15, 16, 29]:

| Stage | Cost | | Weight | | Probability of Survival |
|---|---|---|---|---|---|
| $j$ | $c_j$ | C | $w_j$ | W | $R_j$ |
| 1 | 1.2 | | 1.0 | | 0.20 |
| 2 | 2.3 | | 1.0 | | 0.30 |
| 3 | 3.4 | 47.0 | 1.0 | 20.0 | 0.25 |
| 4 | 4.5 | | 1.0 | | 0.15 |

*Example 11.3.*    (Nonlinear Objective Function and Nonlinear Constraint
Functions):   In this example [17, 25, 26], the system has N stages
operating in series.  We want to achieve a system reliability being
at least $R_{s,min}$ while minimizing the cost.  To attain this reliability,
redundant components, $m_j$, are added in parallel up to a maximum of
an allowed number, $m_{j,max}$, at each stage.  The problem is to:

Minimize

$$Z = \sum_{j=1}^{N} v_j m_j \exp (-m_j/2)$$

subject to

$$g_1 = \sum_{j=1}^{N} p_{j1} m_j + p_{j2} m_j^2 + p_{j3} \leq P$$

$$g_2 = \sum_{j=1}^{N} c_j \cdot [m_j + \exp(-m_j) - \alpha_j] \geq C$$

$$g_3 = \sum_{j=1}^{N} w_j m_j \exp(-m_j/4) \geq W$$

$$R_s = \prod_{j=1}^{N} [1 - (1 - R_j)^{m_j+1}] \geq R_{s,min}$$

$$0 \leq m_j \leq m_{j,max}, \qquad j = 1,2,\ldots,N$$

The constants assigned to this problem are:

$$N = 2, \ R_{s,min} = 0.85, \ m_{j,max} = 4, \ P = 37, \ C = 81, \ W = 38$$

| $j$ | $v_j$ | $p_{j1}$ | $p_{j2}$ | $p_{j3}$ | $c_j$ | $\alpha_j$ | $w_j$ | $R_j$ |
|---|---|---|---|---|---|---|---|---|
| 1 | 3 | 3 | 1 | 0 | 30 | 0 | 30 | .90 |
| 2 | 2 | 3 | 1 | 1 | 30 | 4 | 30 | .75 |

*Example 11.4.* The objective is to maximize the nonlinear system reliability subject to 3 nonlinear constraints with redundant components in each stage subject to type 1 failures [24].

Maximize

$$R_s(m) = \prod_{i=1}^{3} [1 - \sum_{u=1}^{h_i} [1 - (1 - q_{iu})^{m_i+1}] - \sum_{u=h_i+1}^{s_j} (q_{iu})^{m_i+1}]$$

subject to

$$G_1(m) = (m_1 + 3)^2 + (m_2)^2 + (m_3 + 2)^2 \leq 51$$

$$G_2(m) = 20(m_1 + \exp(-m_1)) + 20(m_2 + \exp(-m_2))$$

$$+ 20(m_3 + \exp(-m_3)) \geq 120$$

$$G_3(m) = 20(m_1 \exp(-m_1/4)) + 20(m_2 \exp(-m_2/4))$$

$$+ 20(m_3 \exp(-m_3/4)) \geq 65$$

$$m = (m_1, m_2, m_3), \quad m_i \text{ positive integer for } i = 1,2,3.$$

The subsystems are subject to four failure modes ($s_i = 4$) with one "O" failure ($h_i = 1$) and three "A" failures, for i = 1,2,3. Here the "O" failure mode implies that if "one" component fails the subsystem fails and the "A" failure mode implies that only if "all" components fail, then the subsystem fails. For each subsystem, the failure probability of an element is shown in Table 1.

Table 1   The Type of Failure and its Failure
          Probability for Each Element

| SUBSYSTEM | TYPE OF FAILURE | FAILURE PROBABILITY |
|:---:|:---:|:---:|
| i | u | $q_{iu}$ |
| 1 | O | 0.01 |
|   | A | 0.05 |
|   | A | 0.10 |
|   | A | 0.18 |
|   | O | 0.08 |
|   | A | 0.02 |
| 2 | A | 0.15 |
|   | A | 0.12 |
|   | O | 0.04 |
|   | A | 0.05 |
| 3 | A | 0.20 |
|   | A | 0.10 |

There are at least five methods for solving these kinds of problems using integer programming:  partial enumeration - Lawler and Bell, implicit enumeration - Lemke and Spielberg; cutting plane method - Gomory; Branch and bound - Balas; and implicit enumeration - Geoffrion. They are classified in Table 2.

## 11.2   THE PARTIAL ENUMERATION METHOD

The integer programming method for 0-1 type variables due to Lawler and Bell [11] is used to find the solution of the example. Lawler and Bell describe a programmed algorithm for solving discrete optimization problems with a monotonic objective function and arbitrary constraints.

A brief review of the Lawler-Bell method is provided in this section. The type of problems that can be solved by this method have the following form.  Minimize $g_0(x)$ subject to $r$ constraints of the form

$$g_{i1}(x) - g_{i2}(x) \geq 0, \qquad i = 1, \ldots, m$$

where

$$x = (x_1, x_2, \ldots, x_n) \tag{1}$$

and

$$x_j = 0 \text{ or } 1, \qquad j = 1,\ldots,n$$

Each of the functions in eq. (1) must be monotonically nondecreasing in each of its arguments.  With some ingenuity, many problems can be put in this form.

Vector $x$ is "binary" in the sense that each $x_j$ is either 0 or 1; $x \leq y$ if and only if $x_j \leq y_j$ for $j = 1, \ldots, n$, e.g., $x \leq y$, where $x = (0, 1, 0)$ and $y = (0, 1, 1)$.  This is the "vector partial ordering." There is also the lexicographic or numerical ordering of these vectors obtained by identifying $x$.  Define the integer value $N(x) = x_1 2^{n-1} + x_2 2^{n-2} + \ldots + x_n 2^0$.  Numerical ordering is a refinement of the vector

Table 2    Classification of Examples and Approaches.

| EXAMPLES | METHODS APPLIED TO THE EXAMPLES | REFERENCES |
|---|---|---|
| Example 11.1<br>Min. Linear cost function<br>s.t.<br>$R_s \geq R_{s,min}$<br>Linear weight constraint | Partial enumeration (Lawler & Bell) | 17 |
| | Implicit enumeration (Lemke & Spielberg) | 9 |
| Example 11.2<br>Max. $R_s$<br>s.t.<br>2 linear cost constraints | Cutting plane method (Gomory) | 25, 26 |
| | Branch and bound (Balas) | 4, 15, 16, 18 |
| | Partial enumeration (Lawler & Bell) | 17 |
| | Partial enumeration | 14 |
| | Enumeration (Balas or Glover) | 10 |
| Example 11.3<br>Min. Nonlinear cost function<br>s.t.<br>$R_s \geq R_{s,min}$<br>3 nonlinear constraints | Cutting plane method (Gomory) | 25, 26 |
| | Partial enumeration (Lawler & Bell) | 17 |
| Example 11.4<br>Max. $R_s$ with 2 classes of failure modes<br>s.t.<br>3 nonlinear constraints | Cutting plane method (Gomory) | 24 |
| | Implicit enumeration (Geoffrion) | 8 |

partial ordering, i.e., $x \leq y$ implies $N(x) \leq N(y)$: however, $N(x) \leq N(y)$ does not imply $x \leq y$.

Suppose all binary n-vectors are listed in numerical order, i.e.,

$(0, ..., 0, 0, 0)$
$(0, ..., 0, 0, 1)$
$(0, ..., 0, 1, 0)$
$(0, ..., 0, 1, 1)$
$(0, ..., 1, 0, 0)$      etc.

Immediately following an arbitrary vector x, there may (or may not) be a number of vectors x' with the property that $x \leq x'$. Roughly speaking, these are vectors that differ from x only in that they have 1's in place of one or more of the 'right-most' 0's of x. For example, immediately following $x = (0, 1, 0, 0)$ are $(0, 1, 0, 1)$, $(0, 1, 1, 0)$ and $(0, 1, 1, 1)$, each of which is greater than x in the vector partial ordering.

Let x* denote the first vector following x in the numerical ordering that has the property that $x \nleq x^*$. For any given x, the vector x* is easily calculated using a computer as follows:

Treat x as a binary number:

(1) Subtract 1 from x,

(2) Logically 'or' x and x-1 to obtain x*-1,

(3) Add 1 to obtain x*.

Some examples are given here:

```
    Let x        = 0101100
    (1) x - 1    = 0101011
    (2) x* - 1   = 0101111
    (3) x*       = 0110000
    Let x        = 0101011
    (1)  x - 1   = 0101010
    (2) x* - 1   = 0101011
    (3) x*       = 0101100
    Let x        = 0101000
    (1)  x - 1   = 0100111
    (2) x* - 1   = 0101111
    (3) x*       = 0110000
```

Note that $x^* - 1$ is greater than each of $x$, $x + 1$, ..., $x^* - 2$, in the vector partial ordering.

The method is basically a search method, which starts with $x = \{0, 0, ..., 0\}$ and examines the $2^n$ solution vectors in the numerical ordering described above. Further, the labor of examination is cut down considerably by the rules listed below. As the examination proceeds one can retain the least costly up-to-date solution. If $\hat{x}$ is the solution having the "cost" $g_o(\hat{x})$ and $x$ is the vector being examined, then these steps indicate the following conditions under which certain vectors may be skipped.

1)  Test if $g_o(x) \geq g_0(\hat{x})$. If YES, skip to $x^*$ and repeat the operation otherwise proceed to Step 2.

2)  Examine whether $g_{i1}(x^*-1) - g_{i2}(x) \leq 0$ for $i = 1,...,m$. If YES, proceed to Step 3; otherwise skip to $x^*$ and go to Step 1.

3)  Further, if $g_{i1}(x) - g_{i2}(x) \geq 0$, $i = 1, ..., m$, replace $x$ by $\hat{x}$ and skip to $x^*$; otherwise change $x$ to $x + 1$. In either case further execution is transferred to Step 1.

Lawler and Bell [11] call the above steps of the algorithm skipping rules 1,2,3, respectively. Following the above rules, all the vectors are examined and scanning continues until a vector having the maximum numerical order, viz., $\{1, 1, ..., 1\}$, is found. In case the process has skipped to a vector having a numerical order higher than $\{1, ..., 1\}$, designate this state by "overflow" and terminate the procedure. The least "costly" vector recorded provides the optimum solution. One should not be over-whelmed by the number of trials. In practice the number of vectors to be examined may be quite small. For example, in all 11-variable problems with a total of $2^{11}$-solution vectors, only 42 vectors were examined.

*Solution of Example 11.1.* This example is:

Minimize

$$f = 5m_1 + 8m_2$$

subject to

$$[1 - (1 - 0.91)^{1+m_1}][1 - (1 - 0.96)^{1+m_2}] \geq 0.9903$$

$$9(1 + m_1) + 6(1 + m_2) \leq 40$$

$$(2)$$

or

minimize

$$g_0(m) = 5m_1 + 8m_2$$

subject to

$$g_1(m) = \ln(1 - 0.9^{1+m_1}) + \ln(1 - 0.04^{1+m_2}) + .009747 \geq 0$$

$$g_2(m) = 25 - 9m_1 - 6m_2 \geq 0$$

(3)

Before $m_1$ and $m_2$, the nonnegative integer variables, can be transformed to the variables of the zero-one type, it is necessary to estimate their maximum values. This is done by substituting zero for all variables in the constraints in Problem (3) except the one for which the maximum range is desired. Denote these by $m^*_{ij}$, where subscripts i and j refer to the constraint and stage, respectively. Then $m^u_j = \min \{m^*_{ij}\}$ $(i = 1,\ldots,m)$ is an upper bound for $m_j$.

In Problem (3) the upper limits of $m_1$ and $m_2$ can be found by letting $m_1 = 0$ or $m_2 = 0$ alternatively as follows:

Stage 1:

$$\ln(1 - .09^{1+m_1}) + \ln(1 - .04) + .009747 \geq 0$$

or

$$\ln(1 - .09^{1+m_1}) \geq .031075$$

hence

$$m^*_{11} = M_1 \text{ (where } M_1 \to \infty)$$

$$25 - 9m_1 - 6(0) \geq 0$$

or

$$m_1 \leq 2.78$$

Hence

$$m^*_{21} = 2$$

$$m^u_i = \min \{M_1, 2\}$$

$$= 2$$

Stage 2:

$$\ln (1 - .09) + \ln (1 - .04^{1+m}2) + .009747 \geq 0$$

or

$$\ln(1 - .04^{1+m}2) \geq .08564$$

hence

$$m^*_{12} = M_2 \text{ (where } M_2 \to \infty)$$

$$25 - 9(0) - 6m_2 \geq 0$$

or

$$m_2 \leq 4.17$$

Hence

$$m^*_{22} = 4$$

$$m^u_2 = \min \{M_2, 4\}$$

$$= 4$$

Let

$$m_1 = x_{11} + x_{12}$$

$$m_2 = x_{21} + x_{22} + 2x_{23} \qquad\qquad (4)$$

where $x_{ij}$ is either 0 or 1.

Substituting these expressions into Problem (3), one can obtain the following problem as indicated in Problem (1):

$$g_o(x) = 5x_{11} + 5x_{12} + 8x_{21} + 8x_{22} + 16x_{23}$$

$$g_{11}(x) = \ln(1 - .09^{x_{11}+x_{12}+1}) + \ln(1 - .04^{x_{21}+x_{22}+2x_{23}+1})$$
$$+ .009747$$

$$g_{12}(x) = g_{21}(x) = 0$$

$$g_{22}(x) = 9x_{11} + 9x_{12} + 6x_{21} + 6x_{22} + 12x_{23} - 25$$

Now the problem conforms to the Lawler-Bell algorithm. The solution is determined after examining only twelve vectors out of the 32 generated by the five binary variables of eq. (4). The sequence of examination and the different rules applied are indicated in Table 3. The vector ordering used is also shown, viz., $x \equiv \{x_{23}, x_{12}, x_{22}, x_{11}, x_{21}\}$. There are no definite rules about the ordering of these variables. However, it has been observed for all the problems studied that the variables carrying the least "numerical weights" are assigned the "rightmost" position in the ordering. This is done so that the numerical values of $m_1$ and $m_2$ increase as the examination of the solution vectors x proceeds.

To begin Table 3, we set $x = (0, 0, \ldots, 0)$ and $g_o(\hat{x}) = \infty$ and at the end of the table, the solution is $\hat{x}$, and the minimum cost is $g_0(\hat{x})$.

Iteration 1)

$$x \qquad = (0, 0, 0, 0, 0)$$

$$x^* - 1 = (0, 0, 0, 0, 0)$$

$$g_{11}(x^*-1) - g_{12}(x) = \ln(1 - .09) + \ln(1 - .04) + .009747$$
$$= .125386 < 0$$

Table 3    The Examination Sequence of Example 11.1 by Lawler-Bell
Algorithm.

| $x_{23}$ | $x_{12}$ | $x_{22}$ | $x_{11}$ | $x_{21}$ | | |
|---|---|---|---|---|---|---|
| 0 | 0 | 0 | 0 | 0 | $g_{11}(x^*-1) - g_{12}(x) < 0$ | skip to $x^*$ through step 2 |
| 0 | 0 | 0 | 0 | 1 | $g_{11}(x^*-1) - g_{12}(x) < 0$ | skip to $x^*$ through step 2 |
| 0 | 0 | 0 | 1 | 0 | $g_{11}(x) - g_{12}(x) < 0$ | change $x \to x+1$ through step 3 |
| → 0 | 0 | 0 | 1 | 1 | feasible, $g_o(x) = 13$ | skip to $x^*$ through step 3 |
| 0 | 0 | 1 | 0 | 0 | $g_{11}(x) - g_{12}(x) < 0$ | change $x \to x+1$ through step 3 |
| 0 | 0 | 1 | 0 | 1 | $g_o(x) > g_o(\hat{x})$ | skip to $x^*$ through step 1 |
| 0 | 0 | 1 | 1 | 0 | $g_o(x) = g_o(\hat{x})$ | skip to $x^*$ through step 1 |
| 0 | 1 | 0 | 0 | 0 | $g_{11}(x) - g_{12}(x) < 0$ | change $x \to x+1$ through step 3 |
| 0 | 1 | 0 | 0 | 1 | $g_o(x) = g_o(\hat{x})$ | skip to $x^*$ through step 1 |
| 0 | 1 | 0 | 1 | 0 | $g_{11}(x) - g_{12}(x) < 0$ | change $x \to x+1$ through step 3 |
| 0 | 1 | 0 | 1 | 1 | $g_o(x) > g_o(\hat{x})$ | skip to $x^*$ through step 1 |
| 0 | 1 | 1 | 0 | 0 | $g_o(x) > g_o(\hat{x})$ | skip to $x^*$ through step 1 |
| 1 | 0 | 0 | 0 | 0 | $g_o(x) > g_o(\hat{x})$ | skip to $x^*$ through step 1 |

$x^* = 1(0,0,0,0,0)$;  therefore overflow takes place and we stop.

skip to x* through Step 2.

Iteration 2)

$$x = (0,\ 0,\ 0,\ 0,\ 1)$$

$$g_o(x) = 8$$

hence

$$g_o(x) < g_o(\hat{x})$$

$$x* - 1 = (0,\ 0,\ 0,\ 0,\ 1)$$

$$g_{11}(x*-1) - g_{12}(x) = \ln(1 - .09) + \ln(1 - .04^2) + .009747$$

$$= -.086165 < 0$$

skip to x* through Step 2).

Iteration 3)

$$x = (0,\ 0,\ 0,\ 1,\ 0)$$

$$g_o(x) = 5$$

hence

$$g_o(x) < g_o(\hat{x})$$

$$x* - 1 = (0,\ 0,\ 0,\ 1,\ 1)$$

$$g_{11}(x*-1) - g_{12}(x) = \ln(1 - .09^2) + \ln(1 - .04^2) + .009747$$

$$= .000013 > 0$$

$$g_{21}(x*-1) - g_{22}(x) = -(9 + 6 - 25)$$

$$= 10 > 0$$

change x to x + 1 through Step 3).

Iteration 4)

$$x = (0, \ 0, \ 0, \ 1, \ 1)$$

$$g_o(x) = 5 + 8 = 13$$

hence

$$g_o(x) < g_o(\hat{x})$$

$$x^* - 1 = (0, \ 0, \ 0, \ 1, \ 1)$$

$$g_{11}(x^*-1) - g_{12}(x) = \ln(1 - .09^2) + \ln(1 - .04^2) + .009747$$

$$= .000013 > 0$$

$$g_{21}(x^*-1) - g_{22}(x) = -(9 + 6 - 25)$$

$$= 10 > 0$$

$$g_{11}(x) - g_{12}(x) = .000013 > 0$$

$$g_{21}(x) - g_{22}(x) = 10 > 0$$

Therefore, $g_o(x) = 13$ is a feasible solution.

$$\hat{x} = (0, \ 0, \ 0, \ 1, \ 1)$$

skip to x* through Step 3).

Iteration 5)

$$x = (0, \ 0, \ 1, \ 0, \ 0)$$

$$g_o(x) = 8$$

hence

$$g_o(x) < g_o(\hat{x})$$

$$x^* - 1 = (0, 0, 1, 1, 1)$$

$$g_{11}(x^*-1) - g_{12}(x) = \ln (1 - .09^2) + \ln (1 - .04^3) + .009747$$

$$= .001550 > 0$$

$$g_{21}(x^*-1) - g_{22}(x) = - (6 - 25)$$

$$= 19 > 0$$

$$g_{11}(x) - g_{12}(x) = \ln (1 - .09) + \ln (1 - .04^2) + .009747$$

$$= -.086165 < 0;$$

change x to x + 1 through Step 3).

The process is repeated, and the true optimum is shown by the arrow in Table 3. Therefore, $m_1 = m_2 = 1$ from eq. (4).

Actually, in a large problem there is an appreciable reduction in the number of solution vectors being inspected. For example in a 5-stage problem of Bellman [27] requiring 11 binary variables, the solution was obtained by examining 42 of the $2^{11}$ solutions.

*Solution of Example 11-3.* To solve this problem by the integer programming method of Lawler-Bell, we reformulate the problem as [17]:

Minimize the cost function
$$g_0(m) = [3m_1 \exp (- m_1/2)] + [2m_2 \exp (- m_2/2)]$$
subject to the constraints

$$g_1(m) - 37 - [3m_1 - m_1^2] - [3m_1 + m_2^2 + 1] \geq 0$$
$$g_2(m) = [30(m_1 + \exp (- m_1))] + [30(m_2 + \exp (- m_2)) - 4] - 81 \geq 0$$
$$g_3(m) = [30m_1 \exp (- m_1/4)] + [30m_2 \exp (- m_2/4)] - 38 \geq 0$$
$$g_4(m) = [\ln (1 - 0.1^{m_1+1})] + [\ln (1 - 0.25^{m_2+1})] + 0.1625 \geq 0$$

$$(5)$$

Table 4   The Examination Sequence of Example 11.3 by the Lawler-Bell Algorithm.

| $x_{13}$ | $x_{23}$ | $x_{12}$ | $x_{22}$ | $x_{11}$ | $x_{21}$ | | |
|---|---|---|---|---|---|---|---|
| 0 | 0 | 0 | 0 | 0 | 0 | $g_{21}(x^* - 1) - g_{22}(x) < 0$ | skip to $x^*$ through step 3 |
| 0 | 0 | 0 | 0 | 0 | 1 | $g_{21}(x^* - 1) - g_{22}(x) < 0$ | skip to $x^*$ through step 3 |
| 0 | 0 | 0 | 0 | 1 | 0 | $g_{21}(x^* - 1) - g_{22}(x) < 0$ | skip to $x^*$ through step 3 |
| 0 | 0 | 0 | 2 | 0 | 0 | $g_{31}(x) - g_{32}(x) > 0,\ x \to x + 1$ | |
| 0 | 0 | 0 | 1 | 0 | 1 | feasible, $g_0(\hat{x}) = 1.338780$ | skip to $x^*$ through step 3 |
| 0 | 0 | 0 | 1 | 1 | 0 | $g_0(x) Kg_0(\hat{x})$ | skip to $x^*$ through step 1 |
| 0 | 0 | 1 | 0 | 0 | 0 | $g_0(x) Kg_0(\hat{x})$ | skip to $x^*$ through step 1 |
| → 0 | 1 | 0 | 0 | 0 | 0 | feasible, $g_0(\hat{x}) = 1.082680$ | skip to $x^*$ through step 3 |
| 1 | 0 | 0 | 0 | 0 | 0 | $g_0(x) Kg_0(\hat{x})$ | skip to $x^*$ through step 1 |

$x^* = 1\ (000000)$, i.e., overflow.

$g_4(m)$ refers to the reliability constraint.

The problem can then be transformed to the type (1) problem by substituting $m_1 = x_{11} + 2x_{12} + 4x_{13}$ and $m_2 = x_{21} + 2x_{22} + 4x_{23}$. The maximum value of $m_1$ or $m_2$ is 5 because of the constraints of (5).

The different functions in (1) are defined as follows:

$$g_{11}(x) = g_{22}(x) = g_{32}(x) = g_{42}(x) = 0$$

$$- g_{12}(x) = 36 - 3(x_{11} + 2x_{12} + 4x_{13}) - (x_{11} + 2x_{12} + 4x_{13})^2$$

$$- 3(x_{21} + 2x_{22} + 4x_{23}) - (x_{21} + 2x_{22} + 4x_{23})^2$$

$$g_{21}(x) = 30[(x_{11} + 2x_{12} + 4x_{13}) + \exp(-(x_{11} - 2x_{12} + 4x_{13}))]$$

$$+ 30[(x_{21} + 2x_{22} + 4x_{23}) + \exp(-(x_{21} + 2x_{22} + 4x_{23}))] - 85$$

$$g_{31}(x) = 30[x_{11} + 2x_{12} + 4x_{13}) * \exp(-(\frac{x_{11} + 2x_{12} + 4x_{13}}{4}))]$$

$$+ 30[(x_{21} + 2x_{22} + 4x_{23})(-(\frac{x_{21} + 2x_{22} + 4x_{23}}{4}))] - 38$$

$$g_{41}(x) = \ln(1 - 0.1^{x_{11}+2x_{12}+4x_{13}+1})$$

$$+ \ln(1 - 0.25^{x_{21}+2x_{22}+4x_{23}+1}) + 0.1625$$

The solution with variable ordering is indicated in Table 4 and was obtained in nine steps [17], whereas the complete set has 64 vectors. The minimum $g_0(x)$ recorded is 1.082680 (shown by the arrow in Table 4) for which the allocation is $m_1 = 0$, $m_2 = 4$, $\ln R_s = -0.106563$, and $R_s = 0.89892$.


## 11.3  THE GOMORY CUTTING PLANE METHOD


*Solution of Example 11.2.*  This problem is to be solved by Gomory cutting plane method [5, 6]. The method is applicable to problems whose objective and constraint functions are separable. A separable function

of several variables is one that can be written as a sum of functions
each with only one of the variables as the argument.

The reliability optimization problem can be formally stated as:

$$Z = \sum_{j=i}^{N} f_j(m_j)$$

subject to

$$\sum_{j=1}^{N} g_{ij}(m_j) \leq b_i, \qquad i = 1,2,\ldots,r$$

$$\prod_{j=1}^{N} R_j \geq M$$

\hfill (6)

where

$$m_j = 0,1,\ldots,m_j', \qquad j = 1,2,\ldots,N$$

All terms are known except $Z$ and $m_j$, and $Z$ is the objective to be
maximized, $N$, the number of subsystems or stages, $f_j(m_j)$, the objective
function at state $j$ as a function of $m_j$, $g_{ij}(m_j)$, the amount of the
$i^{th}$ resource consumed at stage $j$ as a function of $m_j$, $b_i$, the amount
of the $i^{th}$ resource available, $r$, the number of constraints, $R_j = 1 -
(1 - R_j)^{m_j+1}$, the reliability of the $j^{th}$ subsystem with $m_j + 1$ units,
where $R_j$ is the reliability of each component, $M$, the minimum acceptable
reliability of the system, and $m_j'$, the maximum number of redundant
units allowed at stage $j$.

After some transformations, the above problem given by eq. (6)
can be solved as the following integer programming problem expressed
by eq. (7):

max/min

$$Z = \sum_{j=1}^{N} \sum_{k=0}^{m_j'} \Delta f_{jk} m_{jk}$$

\hfill (7a)

subject to

$$\sum_{j=1}^{N} \sum_{k=0}^{m'_j} \Delta g_{ijk} m_{jk} \leq b_i, \qquad i = 1, 2, \ldots, r$$

$$\sum_{j=1}^{N} \sum_{k=0}^{m'_j} \Delta \ln R_{jk} m_{jk} \geq \ln M$$

and

$$m_{jk} = 1, \qquad \text{for } k = 0$$

$$m_{jk} - m_{j,k-1} \leq 0, \qquad \begin{array}{l} \text{for } k = 1, \ldots, m'_j \\ \text{for } j = 1, \ldots, N \end{array}$$

$$m_{jk} \geq 0, \qquad \text{for all } j \text{ and } k$$

$$(7b)$$

where in addition to the same notations used in Problem (6), $k$ is index used to denote a particular redundant unit at stage $j$, and $m_{jk}$, the variable representing the $k^{th}$ redundancy at stage $j$, where

$$m_{jk} = 1 \text{ for } 0 \leq k \leq m_j \text{ and } m_{jk} = 0 \text{ for } m_j < k \leq m'_j$$

$$\Delta f_{jk} = f_{jk} \qquad \text{for } k = 0$$

$$= f_{jk} - f_{j,k-1} \qquad \text{for } k = 1, \ldots, m'_j$$

is the change in $f_j(m_j)$ by adding the $k^{th}$ redundancy at stage $j$, where $f_{jk}$ is the objective function of stage $j$ when exactly $k$ redundant units are used.

$$\Delta g_{ijk} = g_{ijk} \qquad \text{for } k = 0$$

$$= g_{ijk} - g_{ij,k-1} \qquad \begin{array}{l} \text{for } k = 1, \ldots, m'_j \\ i = 1, \ldots, r \end{array}$$

is the change in $g_{ij}(m_j)$ by adding the $k^{th}$ redundancy at stage $j$ and where $g_{ijk}$ is the function of the $i^{th}$ resource consumed when $k$ redundant units are used at stage $j$.

$$\Delta \ln R_{jk} = \ln R_{jk} \qquad \text{for } k = 0$$

$$= \ln R_{jk} - \ln R_{j,k-1} \qquad \text{for } k = 1,\ldots,m'_j$$

is the change in $\ln R_j$ by adding the $k^{th}$ redundancy at stage j, and where $R_{jk}$ is the reliability at stage j when k redundant units are used.

The equivalence of Problem (6) and Problem (7) is easily illustrated using the quantities and terms which are defined above and assuming that $m_j$ units are used at stage j.  Substituting these quantities into the objective function Z of (6) yields

$$Z = \sum_{j=1}^{N} f_i(m_j) = \sum_{j=1}^{N} \sum_{k=0}^{m_j} \Delta f_{jk}$$

Likewise the restriction equation of (6) becomes

$$\sum_{j=1}^{N} g_{ij}(m_j) = \sum_{j=1}^{N} \sum_{k=0}^{m_j} \Delta g_{ijk}$$

Now by taking logarithms of the reliability restriction of (6) and with the appropriate substituting, we obtain the equivalent restriction:

$$\ln M \leq \sum_{j=1}^{N} \ln R_j = \sum_{j=1}^{N} \sum_{k=0}^{m_j} \Delta \ln R_{jk}$$

Now the new variable $m_{jk}$ is introduced which represents the $k^{th}$ redundancy at stage j and is defined as:

$$m_{jk} = 1 \qquad \text{for } 0 \leq k \leq m_j$$

$$= 0 \qquad m_j \leq k \leq m'_j \tag{8}$$

with the obvious result that

$$m_j = \sum_{k=1}^{m'_j} m_{jk}$$

In the above, it is understood that the $\Delta \ln R_{jk}$ are numerically evaluated coefficients.  To complete the integer programming formulation it is necessary to formulate the relationships of eq. (8) as restrictions.  Eq. (8) includes the requirement that each subsystem shall contain at least one component.  This is accomplished by including the following restrictions

$$m_{jk} = 1 \qquad \text{for } k = 0$$
$$j = 1,\ldots,N \qquad\qquad (9)$$

The remaining part of eq. (8) insures that at each stage $j$, the $k^{th}$ redundant unit $m_{jk}$ equals one if it is in the solution and that it is in the solution only if the $(k-1)^{th}$ redundant unit is included.  This is incorporated into the problem by including the restraints

$$m_{jk} \leq m_{j,k-1}, \qquad k = 1,\ldots,m_j'$$
$$j = 1,\ldots,N \qquad\qquad (10)$$

The inclusion of eqs. (9) and (10) completes the formulation of Problem (6) as an integer programming problem as stated by (7).  After applying the set A data of Example 11.2 shown in the Introduction section, the problem can be illustrated in Fig. 1 in the required integer programming formulation.  The equations in Group I insure that one basic unit is in each stage.  The Group II equations allow the $k^{th}$ redundant unit to be in the solution only if the $(k-1)^{th}$ redundant unit is included, and require the $m_{jk}$ variables to be either zero or one.  The system restrictions on cost and weight are in Group III.  The $c_j$ equation, representing the $\Delta\ln R_{jk}$ values, is the objective function to be maximized.  This problem was solved by an integer programming algorithm, and the solution is:

$$m_{10} = 1 \qquad m_{20} = 1 \qquad m_{30} = 1 \qquad m_{40} = 1 \qquad m_{50} = 1$$

$$m_{11} = 1 \qquad m_{21} = 1 \qquad m_{31} = 1 \qquad m_{41} = 1 \qquad m_{51} = 1$$

$$m_{12} = 1 \qquad m_{22} = 1 \qquad m_{32} = 1 \qquad m_{42} = 1 \qquad m_{52} = 1$$

$$m_{23} = 1 \qquad m_{33} = 1 \qquad m_{43} = 1$$

$$m_{34} = 1$$

and all other $m_{jk} = 0$.  To summarize, there are

$m_1 = 2$ redundant units at stage 1

$m_2 = 3$ redundant units at stage 2

$m_3 = 4$ redundant units at stage 3

$m_4 = 3$ redundant units at stage 4

$m_5 = 2$ redundant units at stage 5

| Group | $b$ | | Stage I | | | | | Stage II | | | | | Stage III | | | | | Stage IV | | | | | Stage V | | | | |
|---|---|---|---|---|---|---|---|---|---|---|---|---|---|---|---|---|---|---|---|---|---|---|---|---|---|---|---|
| | | | $m_{10}$ | $m_{11}$ | $m_{12}$ | $m_{13}$ | $m_{14}$ | $m_{20}$ | $m_{21}$ | $m_{22}$ | $m_{23}$ | $m_{24}$ | $m_{30}$ | $m_{31}$ | $m_{32}$ | $m_{33}$ | $m_{34}$ | $m_{40}$ | $m_{41}$ | $m_{42}$ | $m_{43}$ | $m_{44}$ | $m_{50}$ | $m_{51}$ | $m_{52}$ | $m_{53}$ | $m_{54}$ |
| Group I | 1 | = | 1 | | | | | | | | | | | | | | | | | | | | | | | | |
| | 1 | = | | | | | | 1 | | | | | | | | | | | | | | | | | | | |
| | 1 | = | | | | | | | | | | | 1 | | | | | | | | | | | | | | |
| | 1 | = | | | | | | | | | | | | | | | | 1 | | | | | | | | | |
| | 1 | = | | | | | | | | | | | | | | | | | | | | | 1 | | | | |
| Group II | 0 | ≥ | -1 | 1 | | | | | | | | | | | | | | | | | | | | | | | |
| | 0 | ≥ | | -1 | 1 | | | | | | | | | | | | | | | | | | | | | | |
| | 0 | ≥ | | | -1 | 1 | | | | | | | | | | | | | | | | | | | | | |
| | 0 | ≥ | | | | -1 | 1 | | | | | | | | | | | | | | | | | | | | |
| | 0 | ≥ | | | | | | -1 | 1 | | | | | | | | | | | | | | | | | | |
| | 0 | ≥ | | | | | | | -1 | 1 | | | | | | | | | | | | | | | | | |
| | 0 | ≥ | | | | | | | | -1 | 1 | | | | | | | | | | | | | | | | |
| | 0 | ≥ | | | | | | | | | -1 | 1 | | | | | | | | | | | | | | | |
| | 0 | ≥ | | | | | | | | | | | -1 | 1 | | | | | | | | | | | | | |
| | 0 | ≥ | | | | | | | | | | | | -1 | 1 | | | | | | | | | | | | |
| | 0 | ≥ | | | | | | | | | | | | | -1 | 1 | | | | | | | | | | | |
| | 0 | ≥ | | | | | | | | | | | | | | -1 | 1 | | | | | | | | | | |
| | 0 | ≥ | | | | | | | | | | | | | | | | -1 | 1 | | | | | | | | |
| | 0 | ≥ | | | | | | | | | | | | | | | | | -1 | 1 | | | | | | | |
| | 0 | ≥ | | | | | | | | | | | | | | | | | | -1 | 1 | | | | | | |
| | 0 | ≥ | | | | | | | | | | | | | | | | | | | -1 | 1 | | | | | |
| | 0 | ≥ | | | | | | | | | | | | | | | | | | | | | -1 | 1 | | | |
| | 0 | ≥ | | | | | | | | | | | | | | | | | | | | | | -1 | 1 | | |
| | 0 | ≥ | | | | | | | | | | | | | | | | | | | | | | | -1 | 1 | |
| | 0 | ≥ | | | | | | | | | | | | | | | | | | | | | | | | -1 | 1 |
| Group III | 100 | ≥ | | 5 | 5 | 5 | 5 | | 4 | 4 | 4 | 4 | | 9 | 9 | 9 | 9 | | 7 | 7 | 7 | 7 | | 7 | 7 | 7 | 7 |
| | 104 | ≥ | | 8 | 8 | 8 | 8 | | 9 | 9 | 9 | 9 | | 6 | 6 | 6 | 6 | | 7 | 7 | 7 | 7 | | 8 | 8 | 8 | 8 |
| $c_j$ | | = | -0.10536 | 0.09531 | 0.00904 | 0.00090 | 0.00009 | -0.28768 | 0.22314 | 0.04879 | 0.01183 | 0.00294 | -0.43078 | 0.30010 | 0.08686 | 0.02870 | 0.00985 | -0.22314 | 0.18232 | 0.03279 | 0.00643 | 0.00128 | -0.16252 | 0.13976 | 0.01938 | 0.00287 | 0.00043 |

Figure 1    The Gomory Cutting Plane Formulation of Example 11.2.

This configuration has a cost of 93 units where the limit is 100 units, and weights 104 units which is equal to the limit.  The system reliability $R_s = 0.985$ or $\ln R_s = -0.015175$.

*Solution of Example 11.3.*  To solve this example by the Gomory Cutting Plane Method, the minimum system reliability requirement,

$$R_s \geq R_{s,min}$$

should be transformed into

$$\ln R_s \geq \ln R_{s,min} = - 0.1625$$

This can be written as

$$- 0.1625 \leq \sum_{j=1}^{2} \sum_{k=0}^{4} \Delta\ln R'_{jk} m_{jk}$$

or

$$0.1625 \geq - \sum_{j=1}^{2} \sum_{k=0}^{4} \Delta\ln R'_{jk} m_{jk}$$

The objective is to determine $m_j$, the number of redundant units at stage j, that minimizes the cost function

$$Z = [3m_1 e^{-m_1/2}] + [2m_2 e^{-m_2/2}]$$

while not violating the system restraints:

$$g_1 = [3m_1 + m_1^2] + [3m_2 + m_2^2 + 1] \leq 37$$

$$g_2 = [30(m_1 + e^{-m_1})] + [30(m_2 + e^{-m_2}) - 4] \geq 81$$

$$g_3 = [20m_1 e^{-m_1/4}] + [20m_2 e^{-m_2/4}] \geq 38$$

$$g_4 = \prod_{j=1}^{2} [1 - (1 - R_j)^{m_j+1}] \geq 0.85$$

Problems with fewer linear constraints such as Example 11.1 can be readily solved by methods presented in [20, 27], but it seems that these methods are inadequate for solving Example 11.3 which includes multiple nonlinear restraints.  The integer programming formulation of this problem is illustrated in Fig. 2.

| | | | Stage I | | | | | Stage II | | | | |
|---|---|---|---|---|---|---|---|---|---|---|---|---|
| | | | $m_{10}$ | $m_{11}$ | $m_{12}$ | $m_{13}$ | $m_{14}$ | $m_{20}$ | $m_{21}$ | $m_{22}$ | $m_{23}$ | $m_{24}$ |
| Group I | 81 | $\leq$ | 30 | 11 | 23 | 27 | 29 | 26 | 11 | 23 | 27 | 29 |
| | 38 | $\leq$ | 0 | 23 | 13 | 6 | 2 | 0 | 23 | 13 | 6 | 2 |
| Group II | 1 | $=$ | 1 | | | | | | | | | |
| | 1 | $=$ | | | | | | 1 | | | | |
| Group III | 0 | $\geq$ | -1 | 1 | | | | | | | | |
| | 0 | $\geq$ | | -1 | 1 | | | | | | | |
| | 0 | $\geq$ | | | -1 | 1 | | | | | | |
| | 0 | $\geq$ | | | | -1 | 1 | | | | | |
| | 0 | $\geq$ | | | | | | -1 | 1 | | | |
| | 0 | $\geq$ | | | | | | | -1 | 1 | | |
| | 0 | $\geq$ | | | | | | | | -1 | 1 | |
| | 0 | $\geq$ | | | | | | | | | -1 | 1 |
| Group IV | 37 | $\geq$ | 0 | 4 | 6 | 8 | 10 | 1 | 4 | 6 | 8 | 10 |
| | 0.1625 | $\geq$ | 0.1054 | -0.0953 | -0.0090 | -0.0009 | -0.0001 | 0.2877 | -0.2231 | -0.0488 | -0.0118 | -0.0029 |
| | $c_j$ | $=$ | 0.0 | -1.81959 | -0.38768 | 0.19911 | 0.38415 | 0.0 | -1.21306 | -0.25846 | 0.13274 | 0.25510 |

Figure 2    The Gomory Cutting Plane Formulation of Example 11.3.

The Group I equations represent the greater than restriction, and the Group IV equations represent the less than restrictions on the system.  The Group II equations insure that one basic unit is in each stage. The Group III equations allow the $k^{th}$ redundant unit to be in the solution only if the $(k-1)^{th}$ redundant unit is included and requires the $m_{jk}$ variables to be either zero or one. This minimization problem is converted to a maximization problem by multiplying the objective function by $(-1)$. The $c_j$ equation is the objective function to be maximized. The integer programming solution is as follows

$$m_{10} = 1 \qquad m_{20} = 1 \qquad m_{21} = 1 \qquad m_{24} = 1$$

$$m_{22} = 1 \qquad m_{23} = 1$$

and all other $m_{jk} = 0$. To summarize, there are

$m_1 = 0$ redundant units at stage 1

$m_2 = 4$ redundant units at stage 2

The minimum cost $Z = 1.0827$ and the system reliability $R_s = 0.899$ where $\ln R_s = -0.1065$.


## 11.4   THE BRANCH AND BOUND METHOD


*Solution of Example 11.2.*   Example 11.2 can also be solved by the branch and bound method [4, 28], which is briefly introduced as follows:

Problem A:   Maximize the total system reliability

$$R_s = \prod_{i=1}^{m} (1 - p_i^{n_i})$$

subject to the constraints:

$$\sum_{i=1}^{m} a_{ij} n_i \le d_j, \qquad j = 1,\ldots,s, \ n_i \ge 1; \ n_i \text{ is integer}$$

If we make the following transformations:

$$c_{ik} = \ln (1 - p_i^{k+1}) - \ln (1 - p_i^{k})$$

$$b_j = d_j - \sum_{i=1}^{m} a_{ij}$$

then Problem A can be identically formulated as:

Problem B: Maximize

$$Z = \sum_{i=1}^{m} \sum_{k=1}^{\infty} c_{ik} x_{ik}$$

subject to constraints:

$$\sum_{i=1}^{m} \sum_{k=1}^{\infty} a_{ij} x_{ik} \leq b_j, \qquad j = 1, \ldots, s$$

where $x_{ik} = 0$ or $1$; and $x_{ik} = 0$ implies $x_{i\ell} = 0$ if $\ell > k$, $x_{ik} = 1$ implies $x_{im} = 1$ if $m < k$.

The one-to-one correspondence of Problem A and Problem B can be easily proved:

Let $X = (x_{ik})$ be a feasible solution to Problem B and let $k_i$ be the largest index such that $x_{ik} = 1$.

Since X is a feasible solution for Problem B, then

$$\sum_{i=1}^{m} \sum_{k=1}^{\infty} a_{ij} x_{ik} \leq b_j$$

$$\sum_{i=1}^{m} a_{ij} k_i \leq d_j - \sum_{i=1}^{m} a_{ij}$$

$$\sum_{i=1}^{m} a_{ij} (k_i + 1) \leq d_j$$

Hence $N = (n_i \mid n_i = k_i + 1)$ is a feasible solution for Problem A. The other constraints are satisfied since $k_i$ is a nonnegative integer.

The objective function for the feasible solution X in Problem B is given by

$$Z = \sum_{i=1}^{m} \sum_{k=1}^{\infty} c_{ik} x_{ik}$$

$$= \sum_{i=1}^{m} \sum_{k=1}^{k_i} \{\ln (1 - p_i^{k+1}) - \ln (1 - p_i^{k})\}$$

$$= \sum_{i=1}^{m} \{\ln (1 - p_i^{k_i+1}) - \ln (1 - p_i)\}$$

$$= \ln R_s - \sum_{i=1}^{m} \ln (1 - p_i)$$

As a conclusion, $R_s$ is maximized when Z is maximum; that is, the optimal solution to Problem B corresponds to the optimal solution to Problem A.

The proposed solution procedure first obtains an optimal solution to Problem B and then converts it into an optimal solution for Problem A by using the relation $n_i = k_i + 1$. Problem B is a multi-dimensional knapsack problem (MDK).

The branch-and-bound procedure consists of two phases: (1) partitioning the space of all feasible solutions into mutually exclusive and exhaustive subsets using a consistent decision rule, and (2) establishing an upper (or lower as the case may be) bound over the objective functions within these subsets. These phases of branching and bounding are repeated until a solution is found that is better than the bounds on all the unexplored subsets. This solution is optimal. A complete description of the branch-and-bound process is given by Balas [28].

In order to develop a bounding procedure for an MDK, consider a single-dimensional knapsack problem.

Maximize

$$\sum_{i=1}^{m} \sum_{k=1}^{\infty} c_{ik} x_{ik}$$

subject to a single constraint

$$\sum_{i=1}^{m} \sum_{k=1}^{\infty} a_{ij} x_{ik} \leq b_j \qquad \text{for a given } j$$

Define the ratios $\gamma_{ik} = c_{ik}/a_{ij}$. Then, for a feasible solution,

$$Z = \sum_{i=1}^{m} \sum_{k=1}^{\infty} c_{ik} x_{ik} = \sum_{i=1}^{m} \sum_{k=1}^{\infty} \gamma_{ik} a_{ij} x_{ik}$$

$$\leq \max_{i,k} [\gamma_{ik}] \sum_{i=1}^{m} \sum_{k=1}^{\infty} a_{ij} x_{ik} \leq \max_{i,k} [\gamma_{ik}] * b_j$$

Also, since

$$\exp(c_{ik}) = (1 - p_i^{k+1})/(1 - p_i^{k}) = 1 + p_i^{k}/(1 + p_i + \cdots + p_i^{k-1})$$

and

$$\exp(c_{i,k+1}) = (1 - p^{k+2})/(1 - p^{k+1}) = 1 + p_i^{k+1}/$$

$$(1 + p_i + \ldots + p_i^{k-1} + p_i^k)$$

it can be seen that $c_{ik} > c_{i,k+1}$, which implies $\gamma_{ik} > \gamma_{i,k+1}$, or

$$\max_{i,k} \{\gamma_{ik}\} = \max_i \{\gamma_{i1}\}. \quad \text{Hence } Z \leq \max_i \{\gamma_{i1}\} * b_j.$$

In the MDK there are s constraints, one for each resource j. Therefore, for any feasible solution for the MDK,

$$Z \leq \max_i \{\gamma_{i1}\} * b_j \qquad \text{for any } j$$

or

$$Z \leq \min_j [\max_i \{\gamma_{i1}\} * b_j]$$

Consequently, the optimal feasible solution $Z^*$ is bounded by the quantity $\min_j [\max_i \{\gamma_{i1}\} * b_j]$. This quantity is the upper bound for the MDK. Let $X = (x_{ik})$ be an intermediate solution in which none of the resources is fully utilized. This intermediate solution can be augmented by including $x_{i\ell}$ and i and $\ell$ satisfy the conditions (1) $x_{i\ell-1} = 1$, (2) $x_{i\ell} = 0$, and (3) no excluded decision has previously been made for $x_{i\ell}$. Any such qualified variable can form the basis of a decision either to include or to exclude. This decision would partition the set of all feasible solutions based on the intermediate solution X into two mutually exclusive and exhaustive subsets, and it would be a basis for branching. The subset described by the decision to include $x_{i\ell}$ (i.e., $x_{i\ell} = 1$) would be termed an inclusive branch, and the subset described by the decision to exclude (i.e., $x_{i\ell} = 0$) would be termed an exclusive branch.

Let $k_i$ be defined, as before, as the largest index such that $x_{ik} = 1$ before the branching decision. It can be seen that $\ell = k_i + 1$. Also, let I be the set of all indices i for which an excluded decision is made before the branching decision. Then the bounds for the inclusive and exclusive branches (subsets) can be computed as follows.

Inclusive branch:

$$k_{i*} = k_i + 1$$

$$\text{Unallocated resource } b_j' = b_j - \sum_{i=1}^{m} \sum_{k=1}^{k_i} a_{ij} x_{ij} = b_j - \sum_{i=1}^{m} k_i * a_{ij}$$

$$\text{Objective function (after branching)} = \sum_{i=1}^{m} \sum_{k=1}^{k_i} c_{ik}$$

Hence the upper bound on the inclusive branch equals

$$\sum_{i=1}^{m} \sum_{k=1}^{k_i} c_{ik} + \min_j \left( \max_{i \notin I} (r_{i,k_i+1}) * b_j' \right) \tag{11}$$

Exclusive branch:

$$I' = I \cup i*$$

$$\text{Unallocated resource } b_j' = b_j - \sum_{i=1}^{m} k_i * a_{ij}$$

$$\text{Objective function (after branching)} = \sum_{i=1}^{m} \sum_{k=1}^{k_i} c_{ik}$$

Hence the upper bound on the exclusive branch equals

$$\sum_{i=1}^{m} \sum_{k=1}^{k_i} c_{ik} + \min_j \left( \max_{i \notin I'} (r_{i,k_i+1}) * b_j' \right] \tag{12}$$

Figure 3 shows the computational flow chart using inclusive or exclusive decisions for branching, and eqs. (11) and (12) for computing the bounds on the objective functions. The first forward solution is obtained by selecting the component for a branching decision that yields the highest upperbound on the inclusive branch. During the

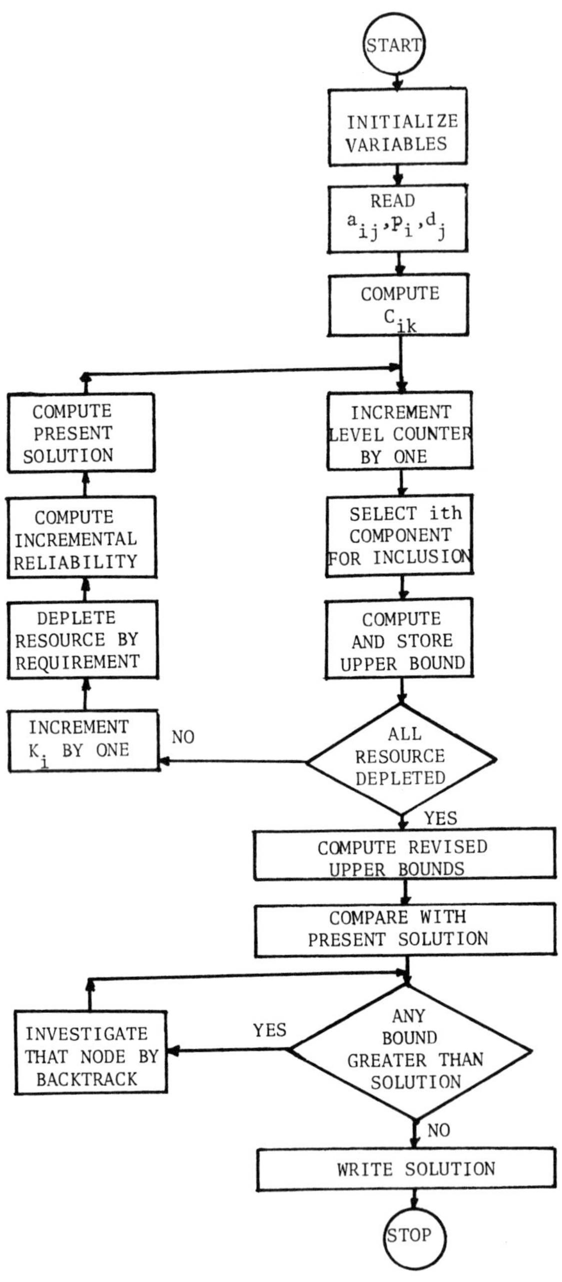



Figure 3   Macro Flowchart of Computational Procedure.

forward procedure, the bounds for the exclusive branch are stored as temporary bounds. The bounds for the inclusive branches are not stored explicitly. After a complete solution is reached (i.e., at least one of the resources is depleted completely at a given solution $X^0$), all the temporary bounds on the exclusive branches are revised. For this revision the index $k_i$ is changed to the largest such that $x_{ik} = 1$ in the solution $X^0$, for $i \neq i^*$. These revised upper bounds are then compared with the objective function $Z(X^0)$. Only those branches need to be explored further for which the upper bound exceeds $Z(X^0)$.

With the data set B, Example 11.2 can be converted into the following problem:

Maximize

$$Z = \sum_{i=1}^{4} \sum_{k=1}^{\infty} c_{ik} x_{ik}$$

subject to

$$\sum_{k=1}^{\infty} (1.2x_{1k} + 2.3x_{2k} + 3.4x_{3k} + 4.5x_{4k})$$

$$\leq 47 - (1.2 + 2.3 + 3.4 + 4.5) = 35.6$$

$$\sum_{k=1}^{\infty} (x_{1k} + x_{2k} + x_{3k} + x_{4k}) \leq 20 - (1 + 1 + 1 + 1) = 16$$

where

$$c_{ik} = \ln (1 - p_i^{k+1}) - \ln (1 - p_i^{k})$$

Hence $b_j = \{35.6, 16\}$, and

$$\{c_{ik}\} = \begin{pmatrix}
i=1 & i=2 & i=3 & i=4 & \\
.182322 & .262364 & .223144 & .139762 & k=1 \\
.032790 & .066939 & .048790 & .061158 & k=2 \\
.006431 & .019238 & .011834 & .002874 & k=3 \\
.001281 & .005700 & .002937 & .000430 & k=4 \\
.000256 & .001704 & .000733 & .000065 & k=5 \\
\vdots & \vdots & \vdots & \vdots & \vdots
\end{pmatrix}$$

The computational procedure of Example 11.2, data set B solved by branch-and-bound method, is shown in Table 5.  The optimal solution is reached at $(k_1, k_2, k_3, k_4) = (4, 5, 3, 2)$ or $(n_1, n_2, n_3, n_4) = (5, 6, 4, 3)$ which gives the system reliability, $R_s = 0.991691$.  The same result has been obtained by Proschan and Bray [29].


## 11.5   THE GEOFFRION IMPLICIT ENUMERATION METHOD

The Geoffrion implicit enumeration method was used to solve the reliability optimization problem with two classes of failure modes [24].  A formulation by 0-1 linear programming [3, 8] is introduced. Notation and nonmenclature:

k-out-of-n: F      the system is failed if and only if at least k out of its n elements are failed.

k-out-of-n: G      the system is good if and only if at least k out of its n elements are good.

$m_i$      number of failure modes in subsystem i.

$s_i$      total number of failure modes in subsystem i.

Class 0 failure modes those of which subsystem i is $1\text{-out-of-}m_j\text{:F}$

$h_i$      number of class 0 failure modes in subsystem $i (u=1,2,\ldots,h_i)$

Class A failure modes those of which subsystem i is $1\text{-out-of-}m_i\text{: G}$

$s_i - h_i$      number of class A failure modes in subsystem i $(u = h_i + 1, h_i + 2,\ldots,s_i)$

$q_{iu}$      probability of failure mode u for each element in subsystem i.

$b_t$      the amount of system resource t available, size of the constraint

T      number of kinds of resources $(t = 1,2,\ldots,T)$

$g_{ti}(m_i)$      subsystem i requires this much of resource t when there are $m_i$ elements in it.

$Q_u^0(m_i),\ Q_u^A(m_i)$      failure probability in subsystem i ($m_i$ elements) for failure mode u of the class 0 or A failure modes.

Table 5    Illustration of the Forward Procedure of Example 11.2 with Data Set B.

| LEVEL | $k_i$ | STAGE SELECTED | $\min_j [\max_{i \notin I'} \{r_{i,k_i+1}\} \cdot b_j']$ | $Z = \sum_{i=1}^{m} \sum_{k=1}^{k_i} c_{ik} x_{ik}$ | UPPER BOUND | $b_j'$ REVISED | COMMENT |
|---|---|---|---|---|---|---|---|
| 1 | 0,0,0,0 | 2 | 4.197824 | 0 | 4.197824 | 33.3, 15 | Inclusive branch |
| 2 | 0,1,0,0 | 3 | 3.347160 | .262364 | 3.609524 | 29.9, 14 | Inclusive branch |
| 3 | 0,1,1,0 | 1 | 2.552508 | .485508 | 3.038016 | 28.7, 13 | Inclusive branch |
| 4 | 1,1,1,0 | 4 | 0.891365 | .667830 | 1.559195 | 24.2, 12 | Inclusive branch |
| 5 | 1,1,1,1 | 2 | 0.704317 | .807592 | 1.511909 | 21.9, 11 | Inclusive branch |
| 6 | 1,2,1,1 | 1 | 0.598418 | .874531 | 1.472949 | 20.7, 10 | Inclusive branch |
| 7 | 2,2,1,1 | 3 | 0.297045 | .907321 | 1.204366 | 17.3, 9 | Inclusive branch |
| 8 | 2,2,2,1 | 4 | 0.235124 | .956111 | 1.191235 | 12.8, 8 | Inclusive branch |
| 9 | 2,2,2,2 | 2 | 0.107059 | 1.017269 | 1.124328 | 10.5, 7 | Inclusive branch |
| 10 | 2,3,2,2 | 1 | 0.056270 | 1.036507 | 1.092777 | 9.3, 6 | Inclusive branch |
| 11 | 3,3,2,2 | 3 | 0.032373 | 1.042938 | 1.075311 | 5.9, 5 | Inclusive branch |
| 12 | 3,3,3,2 | 2 | 0.014620 | 1.054772 | 1.069392 | 3.6, 4 | Inclusive branch |
| 13 | 3,4,3,2 | 1 | 0.003845 | 1.060472 | 1.064317 | 2.4, 3 | Inclusive branch |
| 14 | 4,4,3,2 | 3 | 0.002074 | 1.061753 | 1.063827 | -1, 2 | Exclusive branch |
| 15 | 4,4,3,2 | 2 | 0.001778 | 1.061753 | 1.063531 | .1, 2 | Inclusive branch |
| 16 | 4,5,3,2 | none | – | 1.063457 | – | .1, 2 | Stop here |

Initial solution Z = 1.063457

$Q^0(m_i), Q^A(m_i)$      failure probability in subsystem i ($m_i$ elements) subject to the class 0 or A failure modes.

$Q_i(m_i)$      failure probability in subsystem i ($m_i$ elements)

$R(m)$      reliability of the system when the element allocation is m.

$G_t(m)$      the system requires this much of resource t when the element allocation is m.

$X$      matrix with element $x_{ij}$.

$N$      number of subsystems.

$v_i$      upper bound of $m_i$ elements in subsystem i.

$\gamma_i$      lower bound of $m_i$ elements in subsystem i ($\gamma_i = k$ for k-out-of-$m_i$:G subsystem).

$Q_{iu}(m_i), Q_{iu}^P(m_i)$      probability of failure for subsystem i ($m_i$ elements) by mode u; the modes are separated into classes S & P.


Assumptions:

1. All elements in subsystem i are s-independent with respect to failure mode u; for each i, u combination is taken by itself.

2. The system is 1-out-of-N:F, with respect to its subsystems.

3. The failure modes are a partitioning (mutually exclusive and exhaustive) of the failure event.  Thus failure probabilities for failure modes add to get the total failure probability.

4. Within a subsystem, the elements are 1-out-of-$m_i$:G for some failure modes; and at the same time are 1-out-of-$m_i$:F for the other failure modes.

   We state the original system reliability problem as


*Problem A*

Maximize

$$R_s = \prod_{i=1}^{N} [1 - Q_i(m_i)] \tag{13}$$

subject to

$$G_t(m) = \prod_{i=1}^{N} g_{ti}(m_i) \leq b_t, \qquad t = 1,2,\ldots,T \tag{14}$$

where

$$Q_i(m_i) = Q^0(m_i) + Q^A(m_i)$$

$Q^0(m_i)$ and $Q^A(m_i)$ are the unreliability of subsystem i obtained for class 0 failure modes and for class A failure modes, respectively [24].

To formulate Problem A into a 0-1 linear programming problem, we define the following 0-1 variable:

$$x_{ij} = \begin{cases} 1, & \text{allocate j elements to subsystem i} \\ \\ 0, & \text{otherwise} \end{cases} \tag{15}$$

When we introduce this 0-1 variable the nonlinear systems reliability (Problem A) of the NIP-m problem can be expressed by the following linearized objective function:

$$f(X) = \sum_{i=1}^{N} \sum_{j=\gamma_i}^{v_i} c_{ij} x_{ij} \tag{16}$$

where, for all i and j,

$$c_{1j} \equiv \ell n \left[ 1 - \left\{ \sum_{u=1}^{h_i} Q_{iu}^{P}(j) + \sum_{u=h_i+1}^{s_i} Q_{iu}^{S}(j) \right\} \right]$$

$$Q_{iu}^{P}(j) \equiv 1 - (1 - q_{iu})^{j+1}, \qquad Q_{iu}^{S}(j) \equiv (q_{iu})^{j+1} \tag{17}$$

When we introduce the 0-1 variable into the T nonlinear constraints given by eq. (14), we get

$$g_t(X) = \sum_{i=1}^{N} \sum_{j=\gamma_i}^{v_i} a_{tij} x_{ij} \leq b_t, \qquad t = 1,2,\ldots,T \tag{18}$$

$$a_{tij} \equiv g_{ti}(j) \qquad \text{for all t, i, and j} \tag{19}$$

By definition of the 0-1 variable, eq. (15), we add the following N linear constraints to the constraints, eq. (18):

$$g_{T+i}(X) = 1 - \sum_{j=\gamma_i}^{v_i} x_{ij} \geqq 0, \qquad i = 1,2,\ldots,N \tag{20}$$

By introducing the 0-1 variable, we have therefore reformulated Problem A into a ZOLP problem which maximizes the linear objective function given by eqs. (16) and (17) subject to the T + N linear constraints given by eqs. (18) through (20); this is the ZOLP-m problem. It is proved, as follows, that there is a one-to-one correspondence between the NIP-m and the ZOLP-m proposed here.

First we prove that eqs. (16) and (17) are correct. It is obvious that eq. (17) is necessary. In order to prove that it is sufficient, substitute eq. (17) into eq. (16):

$$f(X) = \sum_{i=1}^{N} \sum_{j=\gamma_i}^{v_i} \ln [1 - \{Q_i^0(j) + Q_i^A(j)\}] x_{ij}$$

$$= \sum_{i=1}^{N} \left\{ \sum_{j\epsilon Z_i^+} \ln [1 - \{Q_i^0(j) + Q_i^Q(j)\}] x_{ij} \right.$$

$$\left. + \sum_{j\epsilon \bar{Z}_i^+} \ln [1 - \{Q_i^0(j) + Q_i^A(j)\}] x_{ij} \right\} \tag{21}$$

where

$$Q_i^0(j) = \sum_{u=1}^{n_i} Q_{iu}^P(j), \qquad Q_i^A(j) = \sum_{u=h_i+1}^{s_i} Q_{iu}^S(j)$$

and

$$Z_i = [j \ : \ \gamma_i, \ \gamma_i+1, \ \ldots, \ v_i]$$

is a set of subsystem i and a direct sum of the $Z_i^+$ and $\bar{Z}_i^+$ (which are partitionings of $Z_i$). Let $X^*$ be a feasible solution to the ZOLP-m problem. Then, eq. (21) is as follows:

$$f(X^*) = \sum_{i=1}^{N} \sum_{j \varepsilon Z_i^+} \ln [1 - \{Q_i^0(j) + Q_i^A(j)\}] \, x_{ij}$$

$$= \sum_{i=1}^{N} \ln [1 - \{Q_i^0(j_i^*) + Q_i^A(j_i^*)\}] = \ln R(j^*) \qquad (22)$$

where

$$j^* = (j_1^*, \, j_2^*, \, \ldots, \, j_n^*)$$

Now we prove that eqs. (18) and (19) are correct. It is obvious that eq. (19) is necessary. In order to prove that it is sufficient, substitute eq. (19) into eq. (18):

$$g_t(X) = \sum_{i=1}^{N} \sum_{j=\gamma_i}^{v_i} g_{ti}(j) \, x_{ij}$$

$$= \sum_{i=1}^{N} \left\{ \sum_{j \varepsilon Z_i^+} t_{ti}(j) \, x_{ij} + \sum_{j \varepsilon \bar{Z}_i^+} g_{ti}(j) \, x_{ij} \right\},$$

$$t = 1, 2, \ldots, \, T \qquad (23)$$

Let $X^*$ be a feasible solution to the ZOLP-m problem, then eq. (23) is

$$g_t(X^*) = \sum_{j=1}^{N} \sum_{j \varepsilon Z_j^+} g_{ti}(j) \, x_{.ij} = \sum_{i=1}^{N} g_{ti}(j_i^*)$$

$$= G_t(j^*) \leq b_t, \qquad t = 1, 2, \ldots, T \qquad (24)$$

*Solution of Example 11.4.* The ZOLP-m problem is to maximize the linear objective function given by eqs. (16) and (17) subject to the linear constraints given by eqs. (18) through (20) for $N = 3$ and $T = 3$, where

## Table 6

---

### Objective Function, $x_{ij}$

| $i$ $j$: | 1 | 2 | 3 | 4 |
|---|---|---|---|---|
| 1 | 24.7 | 23.3 | 38.7 | 50.2 |
| 2 | 123.3 | 243.4 | 332.7 | 417.0 |
| 3 | 26.3 | 112.1 | 161.3 | 204.6 |

### Constraints

| $i$ $j$: | 1 | 2 | 3 | 4 | | 1 | 2 | 3 | 4 |
|---|---|---|---|---|---|---|---|---|---|
| 1 | -16 | -25 | -36 | -49 | | 27.4 | 42.7 | 61.0 | 80.4 |
| 2 | - 1 | - 4 | - 9 | -16 | | ------------same------------ | | | |
| 3 | - 9 | -16 | -25 | -36 | | ------------same------------ | | | |

$\underline{\#1}$, $G_1 \le 51.0$ $\qquad\qquad$ $\underline{\#2}$, $G_2 \le -120.0$

| $i$ | 1 | 2 | 3 | 4 | | 1 | 2 | 3 | 4 |
|---|---|---|---|---|---|---|---|---|---|
| 1 | 15.6 | 24.3 | 28.3 | 29.4 | | - 1 | - 1 | - 1 | - 1 |
| 2 | ----------same---------- | | | | | 0 | 0 | 0 | 0 |
| 3 | ----------same---------- | | | | | ------------same------------ | | | |

$\underline{\#3}$, $G_3 \le -65.0$ $\qquad\qquad$ $\underline{\#4}$, $G_4 \le 1$

| $i$ | 1 | 2 | 3 | 4 | | 1 | 2 | 3 | 4 |
|---|---|---|---|---|---|---|---|---|---|
| 1 | 0 | 0 | 0 | 0 | | 0 | 0 | 0 | 0 |
| 2 | - 1 | - 1 | - 1 | - 1 | | ------------same------------ | | | |
| 3 | 0 | 0 | 0 | 0 | | - 1 | - 1 | - 1 | - 1 |

$\underline{\#5}$, $G_5 \le 1$ $\qquad\qquad$ $\underline{\#6}$, $G_6 \le 1$

---

Table 7

---

<u>Feasible Solutions</u>

| i  j: | 1 | 2 | 3 | 4 | | 1 | 2 | 3 | 4 |
|---|---|---|---|---|---|---|---|---|---|
| 1 | 1 | 0 | 0 | 0 | | 0 | 1 | 0 | 0 |
| 2 | 0 | 0 | 0 | 1 | | 0 | 0 | 0 | 1 |
| 3 | 0 | 1 | 0 | 0 | | 1 | 0 | 0 | 0 |

step 37 — step 41

| i  j: | 1 | 2 | 3 | 4 | | 1 | 2 | 3 | 4 |
|---|---|---|---|---|---|---|---|---|---|
| 1 | 0 | 1 | 0 | 0 | | 0 | 1 | 0 | 0 |
| 2 | 0 | 0 | 1 | 0 | | 1 | 0 | 0 | 0 |
| 3 | 1 | 0 | 0 | 0 | | 0 | 0 | 1 | 0 |

step 61 — step 91

| i  j: | 1 | 2 | 3 | 4 |
|---|---|---|---|---|
| 1 | 0 | 1 | 0 | 0 |
| 2 | 1 | 0 | 0 | 0 |
| 3 | 0 | 0 | 1 | 0 |

Optimal Solution

$$m_1^* = 2, \quad m_2^* = 1, \quad m_3^* = 3.$$

---

the coefficients $a_{tij}$ of eq. (18) for j = 1,2,3,4 are:

$$a_{11j} = (3 + j)^2, \quad a_{12j} = (j)^2, \quad a_{13j} = (2 + j)^2$$

$$a_{2ij} = -20(j + \exp(-j)), \quad a_{3ij} = 20j \exp(-j/4)$$

for i = 1, 2, 3

$$b_1 = 51, \quad b_2 = -120, \quad b_3 = -65$$

The ZOLP-m problem is illustrated in Table 6 in the required ZOLP-m formulations which are 1000 times the coefficient $c_{ij}$ for all i and j of the linear objective function given by eq. (16). The variables in Table 6 can be converted into single subscript variables as follows:

$$x_{1j} = x_j, \quad x_{2j} = x_{4+j}, \quad x_{3j} = x_{8+j}$$

The feasible and optimal solution of the ZOLP-m example are shown in Table 7; the optimal solutions are $x_2 = 1$, $x_5 = 1$, and $x_{11} = 1$.


## REFERENCES


1.  Garfinkel, R. S., and G. L. Nemhauser, *Integer Programming,* New York: Wiley (1972).

2.  Gen, M., H. Okuno, and S. Shinofuji, "An optimizing method in system reliability with failure-modes by implicit enumeration algorithm," *J. of the Operations Research of Japan,* Vol. 19, pp. 99-116 (1976).

3.  Geoffrion, A. M., "An improved implicit enumeration approach for integer programming," RAND Corp. Rept. RM-5644-PR June 1968, or *Operations Research,* Vol. 17, pp. 437-454 (1969).

4.  Ghare, P. M., and R. E., Taylor, "Optimal redundancy for reliability in series system," *Operations Research,* Vol. 17, pp. 838-847 (1969).

5. Gomory, R. E., "All integer programming algorithm," IBM Res. Center Rept. RC-189, Jan. 1960, or in *Industrial Scheduling*, J. F. Muth and G. L. Thompson, (eds), Englewood Cliffs, N. J.: Prentice Hall (1963).

6. Gomory, R. E., "Outline of an algorithm for integer solutions to linear problem," *Bulletin of the American Mathematical Society*, Vol. 64, No. 5, pp. 275-278 (1958).

7. Hu, T. C., *Integer Programming and Network Flows*, Reading, Mass.: Addison-Wesley (1969).

8. Hyun, K. M., "Reliability optimization by 0-1 programming for a system with several failure modes," *IEEE Transactions on Reliability*, Vol. R-24, No. 3, pp. 206-210 (1975).

9. Hwang, C. L., L. T. Fan, F. A. Tillman, and S. Kumar, "Optimization of life support system reliability by an integer programming method," *AIIE Transactions*, Vol. 3, No. 3, pp. 229-238 (1971).

10. Koleasar, P. J., "Linear programming and the reliability of multi-component systems," *Naval Research Logistics Quarterly*, Vol. 15, pp. 317-327 (1967).

11. Lawler, E. L., and M. D. Bell, "A method for solving discrete optimization problems," *Operations Research*, Vol. 14, pp. 1098-1112 (1966).

12. Lawler, E. L. and D. E. Wood, "Branch-and-bound-methods: A survey," *Operations Research*, Vol. 14, pp. 699-719, (1966).

13. Lemke, C. E., and K. Spielberg, "Direct search algorithms for zero-one and mixed integer programming," *Operations Research*, Vol. 15, pp. 892-914 (1967).

14. Luus, R., "Optimization of system reliability by a new nonlinear integer programming procedure," *IEEE Transactions on Reliability*, Vol. R-24, pp. 14-61 (1975).

15. McLeavey, D. W., "Numerical investigation of parallel redundancy in series systems," *Operations Research*, Vol. 22, pp. 1110-1117 (1974).

16. McLeavey, D. W., and J. A. McLeavey, "Optimization of system reliability by branch-and-bound," *IEEE Transactions on Reliability*, Vol. R-25, No. 5, pp 327-329 (1976).

17. Misra K. B., "A method of solving redundancy optimization problems," *IEEE Transactions on Reliability*, Vol. R-20, No. 3, pp. 117-120 (1971).

18.  Misra, K. B., and J. Sharma, "Reliability optimization of a system by zero-one programming," *Microelectronics and Reliability*, Vol. 12, pp. 229-233 (1973).

19.  Mizukami, K., "Optimum redundancy for maximum system reliability by the method of convex and integer programming," *Operations Research*, Vol. 16, pp. 392-406 (1968).

20.  Proschan, F. and T. A. Bray, "Optimum redundancy under multiple constraints," *Operations Research*, Vol. 13, pp. 800-814 (1965).

21.  Salkin, H. M., *Integer Programming*, Reading, Mass.:  Addison-Wesley (1975).

22.  Salklin, H., and Spielberg, K., "Adaptive Binary Programming," IBM Scientific Research Centre, 410 East 62nd St., New York, N. Y., Report No. 320-2951 (March 1968).

23.  Taha, H. A., *Integer Programming - Theory, Applications and Computations*, New York:  Academic Press (1975).

24.  Tillman, F. A., "Optimization by integer programming of constrained reliability problems with several modes of failure," *IEEE Transactions on Reliability*, Vol. R-18, No. 2, pp. 47-53 (1969).

25.  Tillman, F. A., "Integer programming solutions to constrained reliability optimization problems," *Transactions of Twentieth Annual Technical Conference American Society for Quality Control*, paper number 66-174, pp. 676-693 (1966).

26.  Tillman, F. A., and J. M. Liittschwager, "Integer programming formulation of constrained reliability problems," *Management Science*, Vol. 13, No. 11, pp. 887-899 (1967).

27.  Bellman, R. E., and S. E. Dreyfus, "Dynamic programming and the reliability of multi-component devices," *Operations Research*, Vol. 6, pp. 200-206 (1958).

28.  Balas, E., "A note on the branch-and-bound principle," *Operations Research*, Vol. 16, pp. 442-445 (1968).

29.  Proschan, F., and T. A. Bray, "Optimal redundancy under multiple constraints," *Operations Research*, Vol. 13, pp. 800-814 (1965).

CHAPTER 12                    OTHER METHODS APPLIED TO SYSTEMS RELIABILITY
                                              OPTIMIZATION PROBLEMS


## 12.1    INTRODUCTION

In addition to the methods presented in the previous chapters, there
are several other methods that have been used to analyze systems reliability optimization problems.  One simple classical approach [1, 7,
11, 13, 15, 16] is to maximize the system reliability without considering constraints such as "cost".  The parametric method [4] involves
a transformation of the objective function into a simplified form so
that either the method of Lagrange multipliers and the Kuhn-Tucker
conditions or the modified Box's method [3] can be applied to solve
the transformed problem.

Linear programming has sometimes been included in reliability
optimization techniques for solving (a) an optimization problem with
a linear form of non-negative variables subject to a system of linear
inequalities [9, 17], or (b) an original nonlinear optimization problem
having been transformed through separable programming [21, 23], to a
standard linear form which can be solved by linear programming.

Stochastic methods have also been used in reliability problems
to maximize the system reliability subject to cost restraints [10].
This method is based on a stochastic approach in which probability
distributions are used to describe families of redundancy configurations.
Random search techniques [5] and other miscellaneous optimization techniques [6, 8, 12, 14, 19, 20] are sometimes applied to system reliability optimization problems.

Following are illustrations of the classical approach, the parametric method, linear programming, and separable programming.


## 12.2   A CLASSICAL APPROACH


Gordon [7], and Moskowitz and McLean [13] may be the first to use graphical techniques to solve optimum component redundancy problems to achieve system reliability.  Their objective also is to develop a general mathematical solution for the optimum number of redundant elements in a system, while the reliabilities of the individual components are known, but without considering constraints.  The figures which show the overall reliability as a function of complexity, reliability of components, and redundancy are presented so that the optimal solution can be identified from the figures.

Working from a theorem of Albert [1], Lloyd and Lipow [11] introduced an effort function which is necessary to accomplish the system reliability of a series configuration from the present reliability, $R_s$, to a desired higher level, $\bar{R}_s$.  Let $R_1$, $R_2$,...,$R_n$ denote the subsystems reliabilities; the system reliability is given by

$$R_s = \prod_{i=1}^{n} R_i \tag{1}$$

Since $\bar{R}_s > R_s$, it is necessary to increase at least one of the $R_i$'s to the point that the required reliability, $\bar{R}_s$, will be met, in accordance with eq. (1).  To accomplish such an increase, requires a certain amount of effort, which is to be alloted in some way among the subsystems. The desired system reliability is then achieved with minimum effort and is given as follows:

(A)   Order the known reliabilities $R_1$, $R_2$,..., $R_n$ in nondecreasing order (we assume that such an ordering is implicit in the notation) so that

$$R_1 \leq R_2 \leq \ldots \leq R_n \tag{2}$$

(B)   Increase each of the reliabilities $R_1, R_2, \ldots, R_{K_0}$ to the same value $\bar{R}_0$; but do not increase the reliabilities $R_{K_0+1}, \ldots, R_n$. The number $K_0$ is determined as

$$K_0 = \text{maximum value of } j \text{ such that}$$

$$R_j < \left( \frac{\bar{R}_s}{\displaystyle\prod_{i=j+1}^{n+1} R_i} \right)^{1/j} = r_j \text{ (say)} \tag{3}$$

where $R_{n+1} = 1$ by definition.

The number $\bar{R}_0$ is determined as

$$\bar{R}_0 = \left( \frac{\bar{R}_s}{\displaystyle\prod_{j=K_0+1}^{n+1} R_j} \right)^{1/K_0} \tag{4}$$

(C)   It is evident that the system reliability will then be $\bar{R}_s$, since

$$\text{the new reliability} = \bar{R}_0^{K_0} R_{K_0+1} \ldots R_n = \bar{R}_0^{K_0} \prod_{j=K_0+1}^{n+1} R_j \tag{5}$$

and by using eq. (4) we obtain

$$\text{the new reliability} = \bar{R}_s$$

*A Numerical Example:*   Let $(R_1, R_2, R_3, R_4, R_5, R_6) = (0.75, 0.80, 0.87, 0.90, 0.95, 0.99)$, then

$$R_s = \prod_{j=1}^{6} R_j = 0.4418$$

The required value of the system reliability is $\bar{R}_s = 0.53$. Suppose that we did not consider the selection of $K_0$ by eq. (3) but arbitrarily decided to set $K_0 = 1$ and use eq. (4). We would then obtain

$$\bar{R}_0 = \left[ \frac{0.53}{\displaystyle\prod_{j=2}^{6} R_j \times 1} \right]^{1/1} = 0.8996$$

and we would have

$$\bar{R}_s = 0.53 = 0.8996 \times 0.80 \times 0.87 \times 0.90 \times 0.95 \times 0.99$$

as desired. However, the theorem tells us that the effort to increase reliability has not been allotted in an optimum manner; i.e., more effort has been used than is necessary. Rather, we should determine $K_0$ by eq. (3). To do this we calculate the quantities:

$$r_6 = \left(\frac{0.53}{1}\right)^{1/6} = 0.8996$$

which is smaller than $R_6 = 0.99$. Therefore the $6^{th}$ component is sufficient. Similarly

$$r_5 = \left(\frac{0.53}{0.99 \times 1}\right)^{1/5} = 0.8825$$

which is smaller than $R_5 = 0.95$;

$$r_4 = \left(\frac{0.53}{0.99 \times 0.95 \times 1}\right)^{1/4} = 0.8664$$

which is smaller than $R_4 = 0.90$; and

$$r_3 = \left(\frac{0.53}{0.99 \times 0.95 \times 0.90 \times 1}\right) = 0.8551$$

which is also smaller than $R_3 = 0.87$; therefore, components of stages 5, 4 and 3 are sufficient. However,

$$r_2 = \left(\frac{0.53}{0.99 \times 0.95 \times 0.90 \times 0.87 \times 1}\right)^{1/2} = 0.8484$$

which is greater than $R_2 = 0.80$. Therefore the $2^{nd}$ component is not sufficient. Since 2 is the largest subscript j such that $R_j < r_j$, then $K_0 = 2$, which means to achieve the system reliability, $\bar{R}_s = 0.53$, the minimum effort to be allotted is to increase the $1^{st}$ and $2^{nd}$ component from 0.70 and 0.80 to the same level, $\bar{R}_0 = 0.8484$; whereas the rest of the components are left at their original level. The resulting reliability of the entire system is, as required,

$$R_s = 0.53 = (0.8484)^2 \times 0.87 \times 0.90 \times 0.95 \times 0.99$$

The procedure which suggested the reliability apportionment as illustrated in the above example is based on a theorem of Albert [1], and is known as the effort function minimization.

The effort function $G(x, y)$ of a system is defined as the amount of effort required to increase the system reliability, x, to a higher level, y. Any cost, weight, volume, or power demand can be regarded as a special kind of effort function, whether they are mathematically well described or not. Therefore, the cost minimization problem is an effort function minimization problem. The effort function always satisfies the following requirements:

1) $G(x, y) \geq 0$, which means that to increase reliability from a lower x to a higher level y will always require at least zero effort.

2) $G(x, y)$ is nondecreasing in y for a fixed x and is nonincreasing in x for a fixed y: e.g.,

$$G(0.7, 0.8) \leq G(0.7, 0.85)$$
$$G(0.6, 0.8) \geq G(0.7, 0.8)$$

3) If $x \leq y \leq z$, $G(x, y) + G(y, z) = G(x, z)$, which states that the amount of effort needed to increase the reliability from x to z is equal to the sum of the effort to increase the reliability from x to y, then from y to z. Namely $G(x, y)$ is additive.

4) $G(0, y)$ has a derivative $h(y)$ such that $yh(y)$ is increasing in y, $0 < y < 1$.

For an N-stage series system, we denote $R_i$ and $R_s$ as the reliabilities of the $i^{th}$ stage and the system, respectively. If $\bar{R}_s$ is the minimum requirement of the system reliability and $\bar{R}_i$, the optimal $i^{th}$ stage reliability, then we can readily define the effort function minimization problem as

Minimize

$$\sum_{i=1}^{N} G(R_i, \bar{R}_i)$$

subject to

$$\prod_{i=1}^{N} \bar{R}_i \geq \bar{R}_s$$

To solve this optimization problem, $R_i$, $i = 1,2,\ldots,N$, $\bar{R}_s$, and the effort function $G(R_i, \bar{R}_i)$ should be given, then various optimization techniques, e.g., dynamic programming, the method of Lagrange multipliers and the Kuhn-Tucker conditions, GRG, etc., can be applied to determine the optimal solution.


## 12.3   PARAMETRIC METHOD


*Principle and Historical Background.*  The parametric approach was originally used in evaluating system reliability, especially when the number of components in a system was large or the system configuration was complex.  Probability was treated as a point in a Cartesian space and formulas were derived to evaluate the systems reliability by assigning a parametric value to this probability [2].

If the probability of success of any event is x, then the probability of failure is y = 1 - x, thus the parametric quantities $\phi$ and $\theta$ associated with x and y are defined by

$$\phi \equiv \tan \theta = \frac{y}{x} = \frac{y}{1 - y} = \frac{1 - x}{x} \tag{6}$$

By this transformation, the complex system, whether in the form of
bridges, delta-star, or star-delta, can be expressed by the combinations of these parameters assigned in each subsystem.  The systems
reliability is then automatically obtained by reversing eq. (6).

The parametric method is an intermediate step in transforming
the objective function from terms of component reliability to terms
involving the parameters, $\phi$, subject to constraints; therefore, the
objective function formulated in parametric forms can be solved by
any applicable nonlinear programming technique.  The method of Lagrange
multipliers and the Kuhn-Tucker conditions [4] and the modified Box's
method [3] are such techniques.

*Formulation of the Problem.*  The problem formulation by the parametric
approach principally involves the transformation of the objective function by eq. (6).

For an N-stage series configuration, the system reliability is

$$R_s = \prod_{j=1}^{N} R_j' \tag{7}$$

where

$$R_j' = [1 - (1 - R_j)^{x_j}] \tag{8}$$

If the parameters $\phi_s$ and $\phi_j'$ are defined as

$$\phi_s = \frac{1 - R_s}{R_s} \quad \left(\text{or } R_s = \frac{1}{1 + \phi_s}\right) \tag{9}$$

and

$$\phi_j' = \frac{1 - R_j'}{R_j'} \quad \left(\text{or } R_j' = \frac{1}{1 + \phi_j'}\right) \tag{10}$$

respectively, then eq. (7) can be represented as

$$\phi_s + 1 = \prod_{j=1}^{N} (1 + \phi_j') \tag{11}$$

In most reliability studies, we are dealing with components having a relatively high value of $R_j'$. Using this fact, eq. (11) can be expressed as

$$\phi_s \simeq \sum_{j=1}^{N} \phi_j' \tag{12}$$

From the definition as stated in eq. (6), we have

$$\phi_j' = \tan \theta_j' = \frac{Q_j'}{1 - Q_j'} \tag{13}$$

and

$$\phi_j = \tan \theta_j = \frac{Q_j}{1 - Q_j} \tag{14}$$

Then from eqs. (13) and (14), it is easy to find that

$$Q_j' = \frac{1}{1 + \cot \theta_j'} \tag{15}$$

and

$$Q_j = \frac{1}{1 + \cot \theta_j} \tag{16}$$

Since $R_j' = 1 - Q_j'$, and $R_j = 1 - Q_j$, then eq. (8) becomes

$$Q_j' = Q_j^{x_j} \tag{17}$$

Substituting eqs. (15) and (16) into eq. (17), we obtain

$$1 + \cot \theta_j' = (1 + \cot \theta_j)^{x_j} \tag{18}$$

By eqs. (13) and (14), eq. (17) can be expressed in terms of $\phi_j'$ and $\phi_j$ as

$$1 + \frac{1}{\phi_j'} = (1 + \frac{1}{\phi_j})^{x_j}$$

or equivalently

$$\frac{\phi_j' + 1}{\phi_j'} = \left(\frac{\phi_j + 1}{\phi_j}\right)^{x_j}$$

If both $\phi_j'$ and $\phi_j$ are much smaller than 1, then

$$\phi_j' \simeq \phi_j^{x_j}$$

Substituting the above approximation into eq. (12), then we reformulate the objective function as

$$\phi_s = \sum_{j=1}^{N} \phi_j^{x_j} \tag{19}$$

to be minimized subject to the linear constraints

$$\sum_{j=1}^{N} c_j x_j \leq C$$

$$\tag{20}$$

$$\sum_{j=1}^{N} w_j x_j \leq W$$

To solve eqs. (19) and (20), the Lagrange function is introduced as

$$L = \sum_{j=1}^{N} \phi_j^{x_j} + \lambda_1 [\sum_{j=1}^{N} c_j x_j - C] + \lambda_2 [\sum_{j=1}^{N} w_j x_j - W] \tag{21}$$

The Kuhn-Tucker conditions are:

$$\frac{\partial L}{\partial x_j} = \phi_j^{x_j} \ln \phi_j + \lambda_1 c_j + \lambda_2 w_j = 0 \tag{22}$$

$$\lambda_1 [\sum_{j=1}^{N} c_j x_j - C] = 0 \tag{23}$$

$$\lambda_2 [\sum_{j=1}^{N} w_j x_j - W] = 0 \tag{24}$$

From eq. (22), we obtain

$$x_j = \frac{1}{\ln \phi_j} [\ln (a_j \lambda_1 + b_j \lambda_2)], \qquad j = 1,2,\ldots,N \tag{25}$$

where

$$a_j \equiv - \frac{c_j}{\ln \phi_j}$$

$$b_j \equiv - \frac{w_j}{\ln \phi_j}$$

Substituting eq. (25) into eqs. (23) and (24) yields

$$- \sum_{j=1}^{N} a_j [\ln (a_j \lambda_1 + b_j \lambda_2)] = C \tag{26}$$

$$- \sum_{j=1}^{N} b_j [\ln (a_j \lambda_1 + b_j \lambda_2)] = W \tag{27}$$

Simultaneous eqs. (26) and (27) are solved to obtain $\lambda_1$ and $\lambda_2$. Once $\lambda_1$ and $\lambda_2$ are obtained, $x_j$, $j = 1,2,\ldots,N$ can be found from eq. (25).

*A Numerical Example.*  Consider the five stage problem [18]:

Maximize

$$R_s = \sum_{j=1}^{5} [1 - (1 - R_j)^{x_j}]$$

subject to

$$g_1 = \sum_{j=1}^{5} c_j x_j \leq C$$

$$g_2 = \sum_{j=1}^{5} w_j x_j \leq W$$

The constants associated with the problem are:

| Stage $j$ | Cost $c_j$ | C | Weight $w_j$ | W | Reliability $R_j$ |
|---|---|---|---|---|---|
| 1 | 5 | | 8 | | 0.90 |
| 2 | 4 | | 9 | | 0.75 |
| 3 | 9 | 100 | 6 | 104 | 0.65 |
| 4 | 7 | | 7 | | 0.80 |
| 5 | 7 | | 8 | | 0.85 |

The objective function is transformed by eq. (19) to

Minimize

$$\phi_s = \sum_{j=1}^{N} \phi_j^{x_j}$$

where

$$\phi_j = \frac{1 - R_j}{R_j}, \qquad j = 1,2,3,4,5$$

By using the method of Lagrange multipliers and the Kuhn-Tucker conditions, we introduce multipliers $\lambda_1$ and $\lambda_2$ to obtain the solution shown in Table 1.  The results are identical to those in [18].

Table 1   Numerical Result of the Example [2]

| NUMBER OF COMPONENTS AT EACH STAGE | | | | | COST USED | WEIGHT USED | | |
|---|---|---|---|---|---|---|---|---|
| $x_1$ | $x_2$ | $x_3$ | $x_4$ | $x_5$ | $g_1$ | $g_2$ | $R_s$ | $\lambda$ |
| 2 | 3 | 3 | 2 | 2 | 77 | 91 | 0.87529 | 0.005 |
| 2 | 3 | 4 | 3 | 2 | 93 | 104 | 0.93080 | 0.004 |
| 2 | 3 | 4 | 3 | 2 | 93 | 104 | 0.93080 | 0.003* |
| 2 | 3 | 4 | 3 | 3 | 100 | 112 | 0.94901 | 0.002 |

* is the optimal solution; $\lambda = \lambda_1 + \lambda_2$

## 12.4   LINEAR PROGRAMMING

A linear programming problem arises whenever two or more candidates or activities are competing for limited resources and when it can be assumed that all relationships within the problem are linear.

Since the reliability optimization problem usually has a nonlinear objective function and/or nonlinear constraint functions, unless we linearize the objective and/or constraint functions (or we encounter a specific nonlinear case), linear programming is not applicable. In the next section we introduce separable programming, which is a special class of nonlinear programming and is usually suited for the systems optimization problems adaptable to linear programming.  Note that the separable formulation was illustrated in the previous chapter.  A special case of the reliability allocation problem is solved by linear programming and is presented here.  This section is based on Selman and Grisamore [17].

*Problem statement and formulation.* The problem deals with reliability least cost apportionment which states that:  a company has a system to build which is composed of two subsystems in series.  The reliability requirement for the system is 0.90.  The initial evaluation of the two subsystems yields a reliability of 0.85 for subsystem 1 and 0.87 for subsystem 2.  The product of these two subsystems reliabilities is approximately 0.74.  It is clear that both subsystems' reliabilities must be improved to meet the 0.90 reliability requirement. The relative additional program cost for incremental reliability improvement is determined to be in a ratio of 0.3 to 0.7 (normalized) for subsystem 1 and 2, respectively.  The reliability improvement tradeoff factor between subsystems 1 and 2 is 0.9 and 0.1, respectively, i.e., subsystem 1 approaches the constraint at a rate of 0.9 per incremental increase in reliability.  The problem is to minimize the costs to meet the 0.90 reliability requirement.

| Notation | Definition |
|---|---|
| $R_o$ | The initial system reliability (predicted) |
| $R_r$ | The design goal (the required reliability) |
| $R'_j$ | The initial assessed reliability of subsystem $j$ |
| $\Delta\alpha_j$ | The reliability improvement increment of subsystem $j$ |
| $\alpha_j$ | The exponent corresponding to the reliability $R'_j$, $\alpha_j < 1$ if subsystem $j$ is part of a redundant system whose reliability is written as $R'^{\alpha_j}_j$ |
| $N$ | The total number of serially connected subsystems |
| $\ln(x)$ | The natural logarithm of $x$ |
| $K_j$ | The reliability improvement difficulty factor for subsystem $j$ |

$$0 \leq K_j \leq 1$$

$$\sum_{j=1}^{N} K_j = 1$$

$M_i$  Number of structural variables in the $i^{th}$ equation

$C_{ij}$  The reliability improvement tradeoff factor for the $i^{th}$ tradeoff between a subset of $M_1 \leq N$ subsystems

$$0 \leq C_{ij} \leq 1; \quad \sum_{j=1}^{M_i} C_{ij} = 1$$

$\beta_i$  The minimal tradeoff requirement (in terms of the total reliability improvement increment) for the $i^{th}$ tradeoff

The methodology for pointing out areas for design improvement to meet design goals using linear programming is as follows:

If

$$R_o < R_r; \quad R_o = \prod_{j=1}^{N} R_j'^{\alpha_j}$$

then we wish to maximize the function,

$$\sum_{j=1}^{N} K_j \Delta\alpha_j$$

subject to the constraints

$$\sum_{j=1}^{N} \Delta\alpha_j \ln R_j' \leq \ln R_o - \ln R_r$$

$$\sum_{j=1}^{M_i} C_{ij} \Delta\alpha_j \leq \beta_i$$

(some $C_{ij}$ may be 0)

*A Numerical Example:*  we wish to maximize

$$.3 \, \Delta\alpha_1 + .7 \, \Delta\alpha_2$$

which therefore minimizes (since $\Delta\alpha_j$ are negative) additional reliability program costs subject to the constraints. For simplication it is assumed that $\alpha_1 = \alpha_2 = 1$.

|  |  | Type of Constraints |
|---|---|---|
| (1) | $.1625\ \Delta\alpha_1 + .1393\ \Delta\alpha_2 \leq -.1964$ | Reliability requirement constraint |
| (2) | $.9\ \Delta\alpha_1 + .1\ \Delta\alpha_2 \leq \beta$ | Tradeoff constraint |
| (3) | $\Delta\alpha_1 > -1$ | Implied constraint |
| (4) | $\Delta\alpha_2 > -1$ | Implied constraint |

Solutions are generated as a function of $\beta$ in Figs. 1 and 2. One can see graphically the following situations related to the feasibility of solutions and the value of $\beta$.

For $\beta \geq -.4161$, the tradeoff constraint does not influence the external solution to the problem. This case represents situations where there is no problem in meeting a given tradeoff constraint.

For $-.9243 \leq \beta \leq -.4161$, there exists a feasible solution, the solution of which is influenced by both the minimal reliability requirement and the tradeoff constraint.

For $\beta < -.9243$, no feasible solution exists since the constraint $\Delta\alpha_1 > -1$ implies that the boundary is open at the left in Fig. 1.

## 12.5 SEPARABLE PROGRAMMING

Separable programming is a special class of nonlinear programming that is adaptable to linear programming. The problems are constructed of separable functions which have the form

$$\phi(\bar{x}) = \sum_{i=1}^{m} h_i(x_i)$$

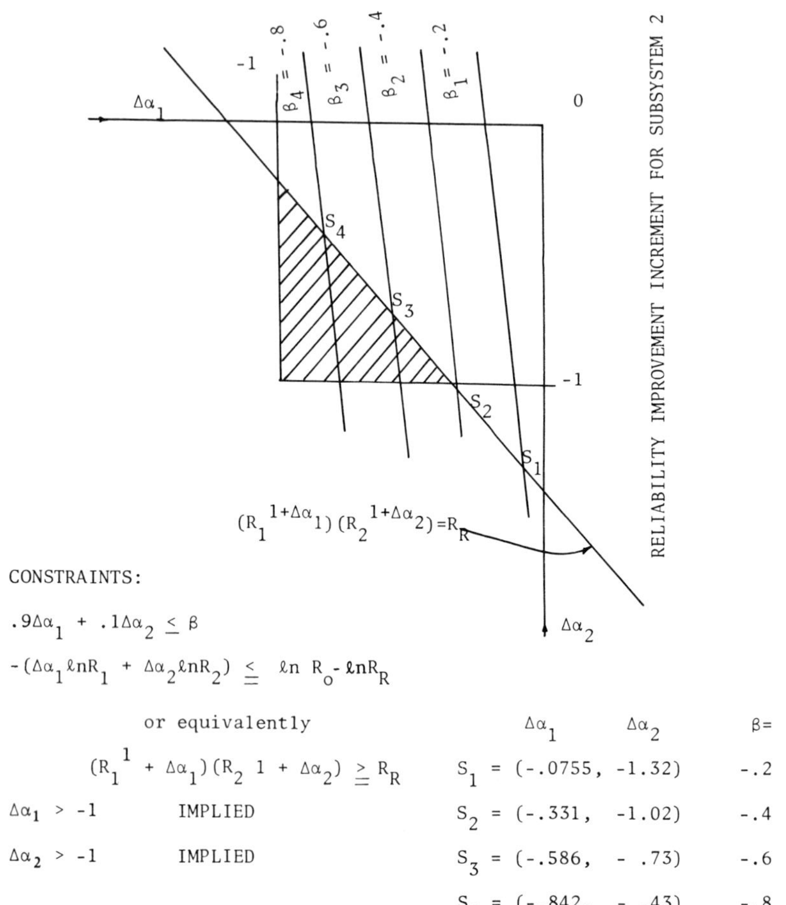


CONSTRAINTS:

$$.9\Delta\alpha_1 + .1\Delta\alpha_2 \leq \beta$$

$$-(\Delta\alpha_1 \ln R_1 + \Delta\alpha_2 \ln R_2) \leq \ln R_o - \ln R_R$$

or equivalently

$$(R_1^{1} + \Delta\alpha_1)(R_2 1 + \Delta\alpha_2) \geq R_R$$

$\Delta\alpha_1 > -1$      IMPLIED

$\Delta\alpha_2 > -1$      IMPLIED

|       | $\Delta\alpha_1$ | $\Delta\alpha_2$ | $\beta=$ |
|-------|----------|----------|------|
| $S_1 =$ | (-.0755, | -1.32)   | -.2  |
| $S_2 =$ | (-.331,  | -1.02)   | -.4  |
| $S_3 =$ | (-.586,  | - .73)   | -.6  |
| $S_4 =$ | (-.842,  | - .43)   | -.8  |

Figure 1   $\Delta\alpha_1$ and $\Delta\alpha_2$ as a function of $\beta$.

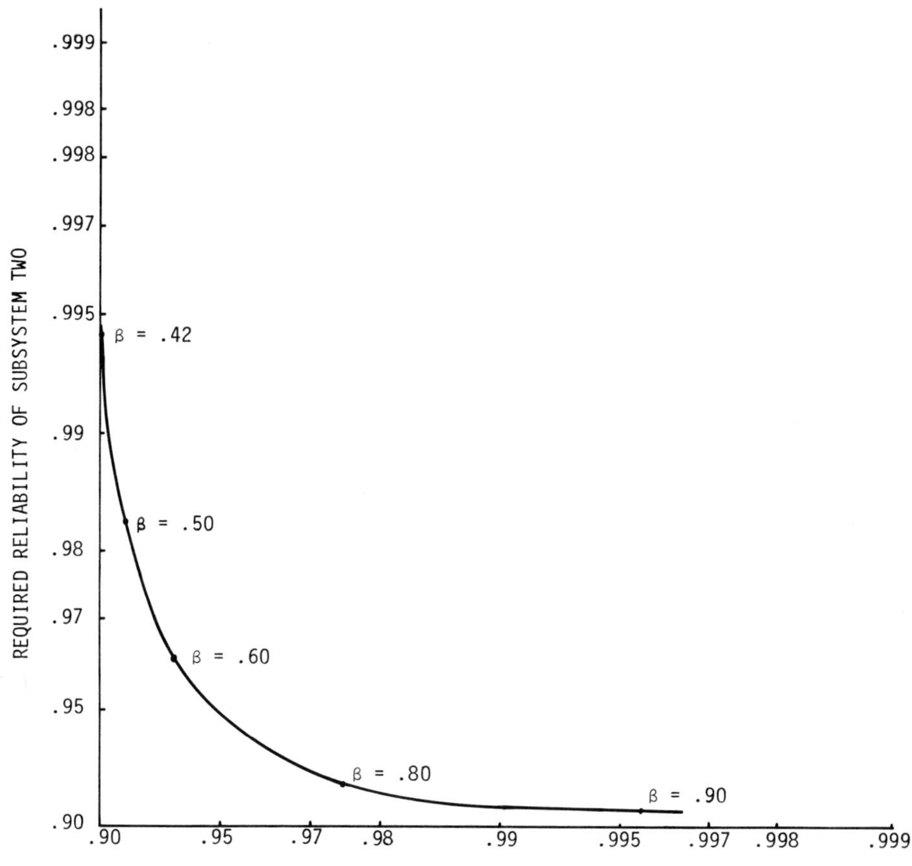


Figure 2   Optimal Values $R_1^{(1 + \Delta\alpha_1)} R_2^{(1 + \Delta\alpha_2)}$ as a function of Beta.

The separable programming problem can be defined as finding a set of $x_i$, $i = 1,2,\ldots,m$ which maximizes (or minimizes)

$$c(\bar{x}) = \sum_{i=1}^{m} f_i(x_i)$$

subject to the constraints

$$\sum_{i=1}^{m} g_{ki}(x_i) \leq b_k, \qquad k = 1,\ldots,p$$

and

$$x_i \geq 0$$

By approximating a nonlinear function of one variable by a piecewise linear function, the problem becomes a restricted linear programming problem, and can be solved by a slightly revised simplex method. This is available on IBM's MPS/360 program [23].

*Formulation of the Problem.*  A continuous nonlinear function of a single variable, $x_i$, can be approximated by a piecewise linear function over a specified interval domain.  This is done by partitioning this interval domain into $n_i$ disjoint, but continuous intervals.  The $(n_i+1)$ points of the partitions are represented by the set

$$S = \left\{ x_i^0, \; x_i^1, \; x_i^2, \; \ldots, \; x_i^{n_i} \right\}$$

There are two methods of representing the piecewise linear approximation of a continuous nonlinear function of one variable.  The method employed here is known as the "delta method."  Both methods are developed in [22].  The "delta method" uses the differences of adjacent points of the set, S, and the differences of the functional values at the adjacent points in developing the approximating equation of a function, $f_i(x_i)$.  The differences are represented by

$$\Delta x_i^j = x_i^j - x_i^{j-1},$$

$$i = 1,2,\ldots,m$$

$$\Delta f_i^j = f_i(x_i^j) - f_i(x_i^{u-1}), \qquad j = 1,2,\ldots,m_i$$

(28)

where the subscript refers to a function and/or variable such as
$x_i$, $f_i(x_i)$, and $g_{ki}(x_i)$, and the superscript refers to a partitioning
of a variable. That is, $f_i(x_i^j)$ is the value of $f_i(x_i)$ at $x_i = x_i^j$.
The differences for adjacent points and the corresponding functional
values for a function with $n_i = 4$ are shown in Fig. 3.

To represent the variable $x_i$ and the approximation of $f_i(x_i)$,
a set of variables, $D_i^j$, $j = 1,2, \ldots, n_i$ is created and follows what
is known as the "restricted-basis-entry-rule." The rule is satisfied
for any one of the following conditions.

$$(i) \quad 0 \leq D_i^1 \leq 1 \quad \text{iff} \quad D_i^j = 0, \quad j = 2,3,\ldots,n_i$$

$$(ii) \quad 0 \leq D_i^j \leq 1 \quad \text{iff} \quad D_i^\ell = 1, \quad \ell = 1,2,\ldots,j-1$$

$$\text{and}$$

$$D_i^k = 0, \quad k = j+1,\ldots,n_i$$

$$(iii) \quad 0 \leq D_i^{n_i} \quad \text{iff} \quad D_i^j = 1, \quad j = 1,2,\ldots,n_i-1$$

where $n_i$ is the number of partitioning intervals for a variable $x_i$.
$D_i^j$ represents a variable created for the $j^{th}$ partition of variable
$x_i$. Intuitively, for any $0 \leq D_i^j \leq 1$ all previous $D_i^\ell$ variables
($\ell = 1,\ldots,j-1$) must have a value of one and all following values
($\ell = j+1,\ldots,n$) must be zero.

*A Numerical Example* [21]: The problem is to

Maximize
$$R_s = \prod_{j=1}^{5} [1 - (1 - R_j)^{x_j}]$$

subject to

$$g_1 = \sum_{j=1}^{5} p_j(x_j)^2 \leq P$$

$$g_2 = \sum_{j=1}^{5} c_j(x_j + \exp(x_j/4)) \leq C$$

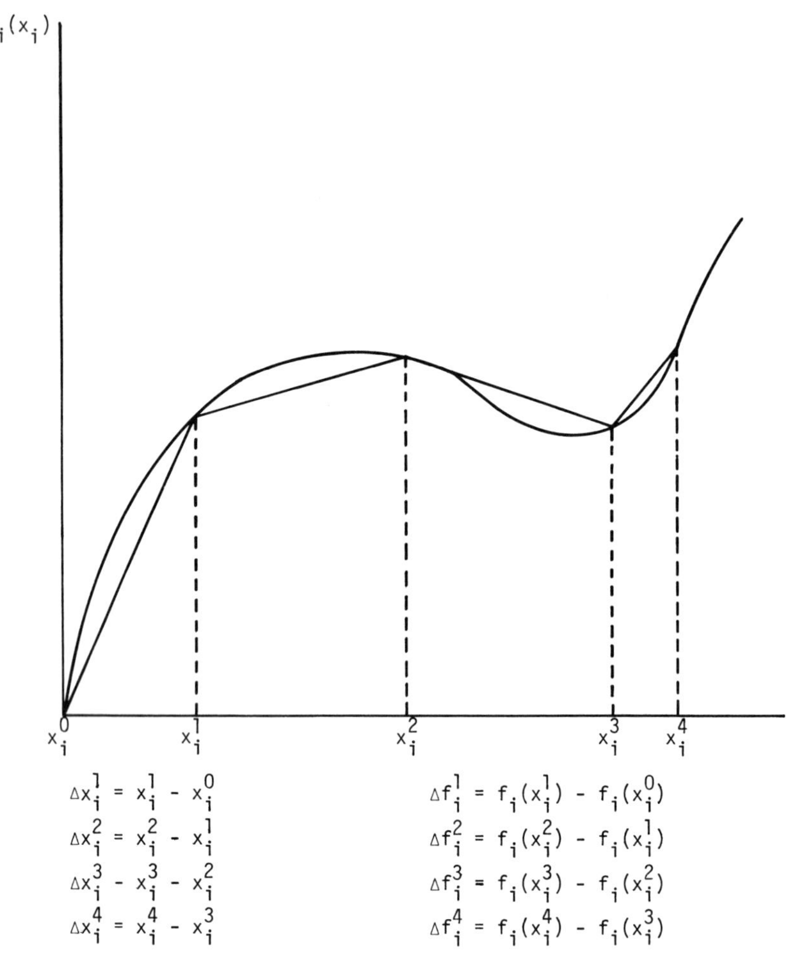



$$\Delta x_i^1 = x_i^1 - x_i^0 \qquad\qquad \Delta f_i^1 = f_i(x_i^1) - f_i(x_i^0)$$

$$\Delta x_i^2 = x_i^2 - x_i^1 \qquad\qquad \Delta f_i^2 = f_i(x_i^2) - f_i(x_i^1)$$

$$\Delta x_i^3 - x_i^3 - x_i^2 \qquad\qquad \Delta f_i^3 = f_i(x_i^3) - f_i(x_i^2)$$

$$\Delta x_i^4 = x_i^4 - x_i^3 \qquad\qquad \Delta f_i^4 = f_i(x_i^4) - f_i(x_i^3)$$

Figure 3   Linear approximation of $f_i(x_i)$.

$$g_3 = \sum_{j=1}^{5} w_j x_j \exp(x_j/4) \leq W$$

where $x_j \geq 1$, $j = 1,2,\ldots,5$ are integers.

The constants associated with the five stage problem are:

| j | $R_j$ | $p_j$ | P | $c_j$ | C | $w_j$ | W |
|---|-------|-------|-----|-------|-----|-------|-----|
| 1 | 0.80 | 1 |     | 7 |     | 7 |     |
| 2 | 0.85 | 2 |     | 7 |     | 8 |     |
| 3 | 0.90 | 3 | 110 | 5 | 175 | 8 | 200 |
| 4 | 0.65 | 4 |     | 9 |     | 6 |     |
| 5 | 0.75 | 2 |     | 4 |     | 9 |     |

It is noted that in optimizing the system reliability problem the
decision variables, namely, the number of components used at each
stage are considered to be continuous variables.  The nearest integer
numbers are assigned eventually.

The objective function is transformed to maximize

$$S = \ln R_s = \sum_{j=1}^{5} \ln [1 - (1 - R_j)^{x_j}]$$

then the problem is formulated into a separable programming problem
and the MPS/360 [23] is applied to solve the problem.

The procedure is recommended in MPS/360 to determine a local optimum solution, if it exists.  Separable programming, at best, will
guarantee only a local optimum.  One reason is that unlike linear inequality constraints, nonlinear inequality constraints do not necessarily form a convex set.  A second reason is that a nonlinear function is not necessarily concave or convex.  The only way to guarantee
a stationary point in a global maximum is for a function to be concave,
or if it is a global minimum, the function must be convex.  Since the
linear approximation function of a separable nonlinear function will
reflect its particular concave and convex properties, separable programming will, at best, produce a local optimum solution.

The solution to the problem is

$$x_1 = 2.70000$$
$$x_2 = 2.32929$$
$$x_3 = 2.10000$$
$$x_4 = 3.50000$$
$$x_5 = 2.80000$$

Following the similar rounding off procedures discussed in Example 8.3 of the GRG chapter, the configuration of (3, 2, 2, 3, 3) will give the optimal solution with a system reliability of, $R_s = 0.9045$, and utilizes $g_1 = 83$, $g_2 = 146.1$ and $g_3 = 194.5$. It is noted that separable programming is an approximate method and its accuracy depends upon the fineness of the grid equations. The uniform grid for this solution is 0.10. The effects of grid size on problem accuracy is dependent on the properties of the approximated functions.


## REFERENCES


1.  Albert, A., "A Measure of the Effort Required to Increase Reliability," Technical Report No. 43,  Applied Mathematics and Statistics Laboratory, Stanford University, Contract No. N6onr-25140 (NR 342-022) (1958).

2.  Banerjee, S. K., and K. Rajamani, "Parametric representation of probability in two dimensions – a new approach in system reliability evaluation," *IEEE Transactions on Reliability,* Vol. R-21, pp. 56-60 (1972).

3.  Banerjee, S. K., K. Rajamani, and S. S. Dishpande, "Optimal redundancy allocation for non series-parallel networks," *IEEE Transactions on Reliability,* Vol. R-25, No. 2, pp. 115-117 (1976).

4.  Banerjee, S. K., and K. Rajamani, "Optimization of system reliability using a parametric approach," *IEEE Transactions on Reliability,* Vol. R-22, pp. 35-39 (1973).

5.  Beraha, D. and K. B. Misra, "Reliability optimization through random search algorithm," *Microelectronics and Reliability,* Vol. 13, pp. 295-297 (1974).

6.   Bodin, L. D., "Optimization procedure for the analysis of coherent structures," *IEEE Transactions on Reliability,* Vol. R-18, No. 3, pp. 118-126 (1969).

7.   Gordon, R., "Optimum component redundancy for maximum system reliability," *Operations Research,* Vol. 5, pp. 229-243 (1957).

8.   Hees, Ir. R. N. V., and Ir. H. W. v. d. Meerendon, "Optimal reliability of parallel multi-component systems," *Operations Research Quarterly,* Vol. 12, No. 1, pp. 16-26 (1961).

9.   Kolesar, P. J., "Linear programming and the reliability of multi-component systems," *Naval Research Logistics Quarterly,* Vol. 15, pp. 317-327 (1967).

10.  Lientz, B. P., "A stochastic method of allocation of components to maximize reliability," *IEEE Transactions on Reliability,* Vol. R-23, No. 2, pp. 91-97 (1974).

11.  Lloyd, D. K., and M. Lipow, *Reliability:  Management, Methods, and Mathematics,* Englewood Cliffs, N. J.: Prentice-Hall (1962).

12.  Morrison, D. F., "The optimum allocation of spare components in system," *Technometrics,* Vol. 3, No. 3, pp. 399-406 (1961).

13.  Moskowitz, F., and J. B. McLean, "Some reliability aspects of system design," *IRE Transactions on Reliability and Quality Control,* Vol. PGROC-8, pp. 7-35 (1956).

14.  Neuner, G. E., and R. N. Miller, "Resource allocation for maximum reliability," *Proceedings of 1966 Annual Symposium on Reliability,* pp. 332-346 (1966).

15.  Sasaki, M., "A simplified method of obtaining highest system reliability," *Proceedings of the Eighth National Symposium on Reliability and Quality Control,* pp. 489-502 (1962).

16.  Sasaki, M., "An easy allotment method achieving maximum system reliability," *Proceedings of the Ninth National Symposium on Reliability and Quality Control,* pp. 109-124 (1963).

17.  Selman, V., and N. T. Grisamore, "Optimum system analysis by linear programming," *1966 Processing Annual Symposium on Reliability,* American Society for Quality Control, pp. 696-703 (1966).

18.  Tillman, F. A., and J. M. Liittschwager, "Integer programming formulation of constrained reliability problems," *Management Science,* Vol. 13, No. 11, pp. 887-899 (1967).

19.  Webster, K. R., "Optimum system reliability and cost effectiveness," *Proceedings of 1967 Annual Symposium on Reliability,* pp. 489-500 (1967).

20.  Webster, L. R., "Choosing optimum system configurations," *Proceedings of the Tenth National Symposium on Reliability & Quality Control*, Washington, D.C., pp. 345-359 (1964).

21.  Williams, J., "Optimization of industrial systems with the separable programming and the generalized reduced gradient methods," M.S. Thesis, Kansas State University, Manhattan, Kansas (1972).

22.  Hadley, G., *Nonlinear and Dynamic Programming*, Reading, Mass: Addison-Wesley (1964).

23.  Mathematical Programming System/360, Version 2, Lineaɪ and Separable Programming - User's Manual, GH20-0476-2, International Business Machines Corp. (1968).

24.  Kapur, K. C., and L. R. Lamberson, Reliability in Engineering Design, New York:  Wiley (1977).

25.  Smith, C. O., *Introduction to Reliability in Design*, New York: McGraw-Hill (1976).

CHAPTER 13   DETERMINATION OF COMPONENT RELIABILITY AND REDUNDANCY
             FOR OPTIMUM SYSTEMS RELIABILITY

## 13.1  INTRODUCTION

In the design process, a system must not only be designed to meet
its functional requirement but must also be designed to perform its
function successfully.  This latter requirement involves designing
reliability into the system.  Often this involves designing to meet
the reliability requirements within the framework of several system
constraints.  In some optimum system reliability problems the element
reliability is assumed to be fixed, and the optimal number of redundancies at each stage is determined where the system is subject to
constraints. A number of optimization techniques have been successfully
applied to solve this class of problems [7, see Chapter 2].  However,
a more general problem is one where both the optimal component reliability and the optimum number of redundancies are to be determined
in order to obtain the best overall systems reliability [3].  Spec-
ifically the problem is one where the designer must determine not
only the number of redundancies but also the reliability of each component.  This is a mixed integer nonlinear programming problem.

In general, problems of this type are difficult to solve by the
normal system optimization techniques, for example, by the method of
Lagrange multipliers [3], sequential unconstrained minimization technique (SUMT) or generalized reduced gradient technique [7] because
these techniques do not provide integer solutions.  The available integer programming techniques do not guarantee an optimal solution.

Hence a technique that provides an integer solution as well as the optimal level of component reliability is required.  The suggested procedure in this chapter is one such technique.

A series system with active component redundancy is considered in this study.  A combination of the well-known Hooke and Jeeves pattern search [2] and the suggested heuristic approach by Aggarwal, et al. [1] is proposed as a stepwise optimization technique for solving this problem.  The procedure is simple and efficient; a component reliability is assumed and the optimal number of redundancies is determined by the heuristic technique.  A sequential search routine for maximizing the overall system reliability is carried out by using the Hooke and Jeeves pattern search.

*Notation:*

$b_i$ = the available resource for the $i^{th}$ constraint

$C$ = the available cost limitation in dollars

$C_j(R_j)$ = the cost of one element at the $j^{th}$ stage as a function of $R_j$

$g_{ij}$ = the amount of the $i^{th}$ resource consumed at the $j^{th}$ stage

$p_j$ = the product of the weight per element and the volume per element at the $j^{th}$ stage

$P$ = the limitation of the product of the volume times the weight constraints

$N$ = the total number of stages in the system of interest

$R_j$ = the initial component reliability at the $j^{th}$ stage

$R_j, Q_j$ = the reliability and unreliability of one element at the $j^{th}$ stage, respectively

$R_s, Q_s$ = the system reliability and unreliability, respectively

$r$ = the total number of constraints

$v_j$ = the volume of one component at the $j^{th}$ stage

$w_i$ = the weight of one component at the $j^{th}$ stage

$W$ = the limitation of weight

$X_j$ = the number of components used at stage $j$

$X_j^o$ = the initial number of elements used at the $j^{th}$ stage

$\bar{X}*(\bar{R})$   = a vector of optimal number of element at each stage as a function of the component reliability at each stage

$\lambda_j$   = the component failure rate at the $j^{th}$ stage

k-out-of-n:F = the system is failed if and only if at least k of its n elements are failed

k-out-of-n:G = the system is good if and only if at least k of its n elements are good


## 13.2   STATEMENT OF THE PROBLEM

The system reliability of an N-stage parallel-series system, where both the component reliability, $R_j$, and the number of components, $X_j$, at the $j^{th}$ stage are to be determined, is expressed by

$$R_s(\bar{R}, \bar{X}) = \prod_{j=1}^{N} [1 - (1 - R_j)^{X_j}] \tag{1}$$

subject to

$$\sum_{j=1}^{N} g_{ij}(R_j, X_j) \le b_i, \qquad i = 1, 2, \ldots, r \tag{2}$$

where the system reliability, $R_s = R_s(R_1, R_2, \ldots, R_N; X_1, X_2, \ldots, X_N)$, $R_j$, $j = 1, 2, \ldots, N$ are all real numbers between 0 and 1, and $X_j$, $j = 1, 2, \ldots, N$ are all positive integers.

To set up eqs. (1) and (2), five assumptions are made:

(1) Each stage is in series and is considered to be essential for the overall operational success of the mission of the system. (The system is denoted as a 1-out-of-N:  F configuration).

(2) All the stages as well as all the parallel elements used at each stage are s-independent.  All components in parallel in the same stage have the same probability of failure.

(3) All the components at each stage are simultaneously working, and for a stage to fail all the elements in that stage must fail. (Each stage is denoted as a 1-out-of-$X_j$:  G configuration).

(4) A short circuit failure will not be considered, that is, only a
single mode of failure is assumed.

(5) The costs are additive between stages.

Both the number of redundancies and the component reliability
improvement will incur a "cost", which may be stated in dollars, weight,
volume or a combination of all three.  In order to be specific, three
such constraints are assumed.  These constraints have been used often
to test and demonstrate optimization techniques [4,5,6].

The first constraint is a combination of weight and volume and
is stated as follows:

$$\sum_{j=1}^{N} g_{1j}(X_j) = \sum_{j=1}^{N} w_j v_j (X_j)^2 + \sum_{j=1}^{N} P_j (X_j)^2 \leq P \tag{3}$$

It is noted that the component reliability does not usually affect
the weight nor the volume, hence $g_{1j}$ is not a function of $R_j$.  The
second constraint is expressed in dollars, and is a function of $X_j$
and $R_j$.  It is stated as:

$$\sum_{j=1}^{N} g_{2j}(X_j, R_j) = \sum_{j=1}^{N} C_j(R_j)(X_j + \exp(X_j/4)) \leq C \tag{4}$$

where $C_j(R_j)$ is the cost per component at the $j^{th}$ stage.  The cost
is an increasing function of $R_j$ or conversely a decreasing function
of the component failure rate expressed by

$$C_j(\lambda_j) = \alpha_j \left\{\frac{1}{\lambda_j}\right\}^{\beta_j}$$

where $\alpha_j$ and $\beta_j$ are constants representing the inherent characteristics of each component at the $j^{th}$ stage, $\beta_j > 1$.  If each component
follows the negative exponential failure law, i.e.,

$$R_j = e^{-\lambda_j t}$$

for all j, then the component cost at the $j^{th}$ stage is

$$C_j(R_j) = \alpha_j \left(\frac{-t}{\ln R_j}\right)^{\beta_j} \tag{5}$$

where t is the operating time during which the component at stage j will not fail. Usually $\alpha_j$ and $\beta_j$ and t are given.

Thus, $C_j(R_j) \cdot X_j$ is the cost of the components at the $j^{th}$ stage as a function of $R_j$ and $X_j$. An additional cost $C_j(R_j) \exp(X_j/4)$ is included, as the cost for interconnecting parallel elements.

Substituting eq. (5) into eq. (4), one obtains a dollar constraint as

$$\sum_{j=1}^{N} \alpha_j \left(\frac{-t}{\ln R_j}\right)^{\beta_j} (X_j + \exp(X_j/4)) \leq C \tag{6}$$

Similarly a weight constraint is stated as

$$\sum_{j=1}^{N} g_{3j}(X_j) = \sum_{j=1}^{N} w_j X_j \exp(X_j/4) \leq W \tag{7}$$

where $w_j X_j$ is the weight of all of the components at the $j^{th}$ stage. Again an additional factor is multiplier, which is $\exp(X_j/4)$, due to the hardware for interconnecting the links. Also note that the weight constraint is not a function of the component reliability.

Now, the problem can be stated as one where the $R_1$, $R_2$, ..., $R_N$; $X_1$, $X_2$, ..., $X_N$ are selected so that eq. (1) will be maximized subject to (3), (6) and (7), where $R_1$, $R_2$, ..., $R_n$ are real numbers between 0 and 1; and $X_1$, $X_2$, ..., $X_N$ are positive integers.

## 13.3   AN OPTIMIZATION PROCEDURE

The combination of the Hooke and Jeeves pattern search [2] and the heuristic approach of Aggarwal, et al. [1] is employed for solving the previously stated mixed integer nonlinear programming problem. The descriptive flow diagram is shown in Fig. 1.

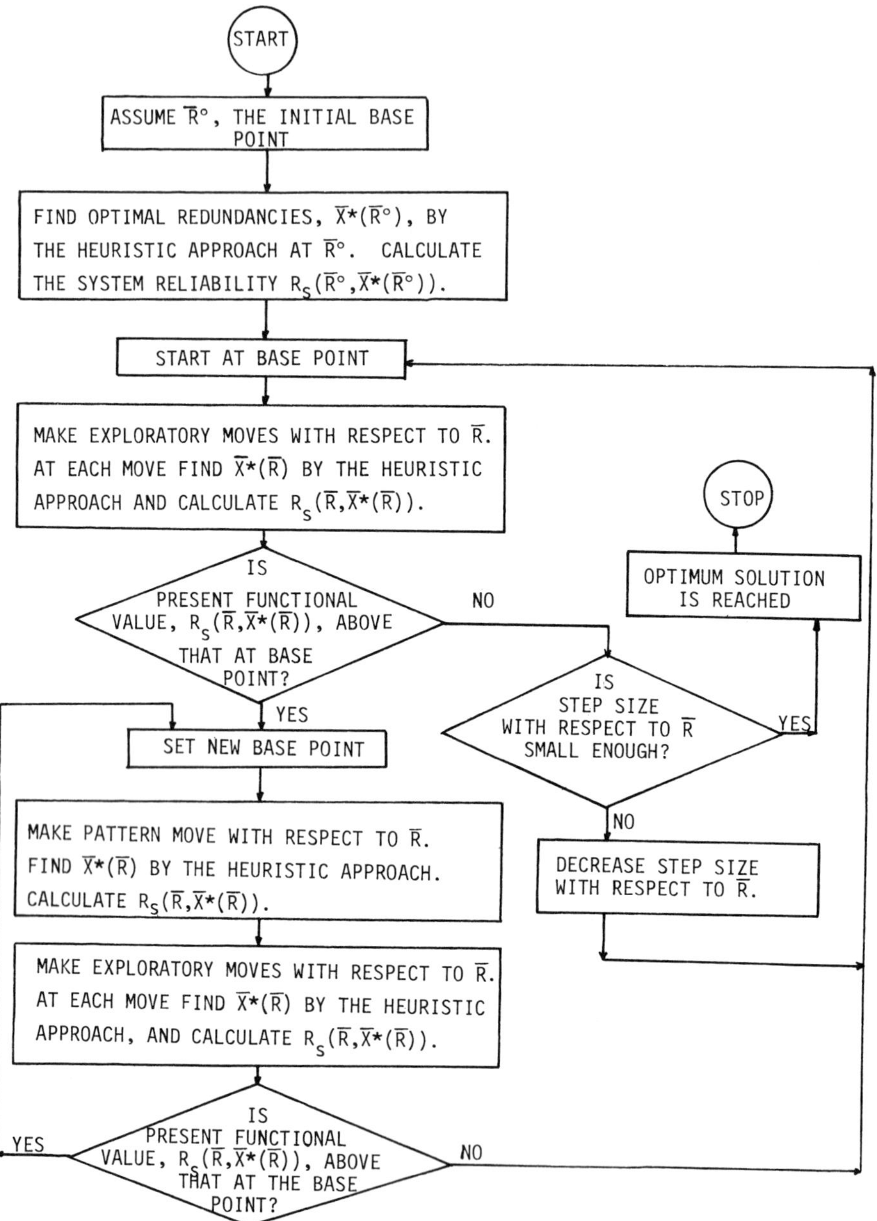



Figure 1     Descriptive Flow Diagram for Combination of Hooke and Jeeves Pattern Search and Heuristic Approach.

The Hooke and Jeeves pattern search technique is a sequential
search routine for maximizing the function, $R_s(\bar{R}, \bar{X})$. The argument
in the Hooke and Jeeves pattern search is the component reliability,
$\bar{R}$, which is varied until the maximum of $R_s(\bar{R}, \bar{X})$ is obtained. The
heuristic approach is applied to each value of $\bar{R}$ to obtain the optimal
number of redundancies, $X_1$, $X_2$, ..., $X_N$, which maximizes $R_s(\bar{R}, \bar{X})$
while satisfying the nonlinear constraints. This heuristic approach
is based on the concept that a component is added to the stage where
its addition produces the greatest ratio of "increment increases in
reliability" to the "product of decrements in slacks". This ratio
is defined by (see Sec. 3.3 of Chapter 3)

$$F_j(X_j) = \frac{\Delta(1 - R_j)^{X_j}}{\prod_{i=1}^{3} \Delta g_{ij}(X_j)} \tag{8}$$

where

$$\Delta(1 - R_j)^{X_j} = (1 - R_j)^{X_j} - (1 - R_j)^{X_j+1} = R_j(1 - R_j)^{X_j}$$

and

$$\Delta g_{ij}(X_j) = g_{ij}(X_j + 1) - g_{ij}(X_j)$$

The computational procedures for evaluating the functional value
of the system reliability, $R_s(\bar{R}, \bar{X})$, at any point is:

1.  For an initial starting point the component reliability,
    $\bar{R} = (R_1, R_2, ..., R_N)$, is given.
2.  (a) Substitute the value $(R_1, R_2, ..., R_N)$ into eqs. (1) and (6),
    then the problem is to find $(X_1, X_2, ..., X_N)$, a straightforward
    redundancy problem where the heuristic approach can be applied.
    (b) Let $\bar{X} = (X_1, X_2, ..., X_N) = (1, 1, ..., 1)$.
3.  (a) Calculate $F_j(X_j)$ for all j using eq. (8).
    (b) Select the stage having the highest $F_j(X_j)$. A redundant com-
    ponent is proposed to be added to that stage.

4.   Check to see if the constraints are violated.

   (a)   If the solution is still feasible, add one redundant component.   Modify the value of $X_j$ and repeat Step 3.

   (b)   If at least one constraint is exactly satisfied; the current
   value of $\bar{X}$ is an optimal solution corresponding to $(R_1, R_2, \ldots, R_N)$.   Go to Step 5.

   (c)   If at least one constraint is violated, cancel the proposed
   addition of the redundant component; remove that stage from further consideration, and repeat Step 3.   When all the stages are
   excluded from further consideration, the current values of $\bar{X}$ are
   the optimal solution with respect to $\bar{R} = (R_1, R_2, \ldots, R_N)$.

5.   Calculate the system reliability, $R_s$, the functional value, for
   the assigned $\bar{R}$ and the optimum $\bar{X}^*$.


## 13.4   NUMERICAL EXAMPLES


*Example 13.1.*   A five-stage problem was solved with the values given
in Table 1.   The optimal solution is presented in Table 2.   The optimum system reliability is 0.91494 at the point $(R_1, R_2, R_3, R_4, R_5;$
$X_1, X_2, X_3, X_4, X_5) = (0.7582, 0.8000, 0.9000, 0.8000, 0.7500; 3, 3,$
2, 2, 3).   Using the starting values of $(R_1, R_2, R_3, R_4, R_5) = (0.70,$
0.70, 0.70, 0.70, 0.70), the computation took 23 sec. to reach the
optimum solution on an IBM 370/158 computer.


*Example 13.2.*   A similar five-stage problem as Example 13.1 was solved, but where the limitations on the constraints were P = 220, C =
350, W = 400.   The optimal solution obtained from using the following
two sets of starting values of $\bar{R}^O = (0.7, 0.7, 0.7, 0.7, 0.7)$ and
$\bar{R}^O = (0.8, 0.8, 0.8, 0.8, 0.8)$ are presented in Tables 3 and 4.   The
optimal system reliabilities for these two set of solutions are
0.993657 and 0.994767.   The difference is about 0.11%. However, the
optimum component reliabilities and redundancies are (0.900, 0.850,
0.856, 0.750, 0.850, 3, 4, 4, 4, 4) and (0.850, 0.863, 0.902, 0.700,
0.900; 4, 4, 3, 4, 3), respectively.   It seems that the functional

Table 1  Constants Used in Example 13.1.

| j | $\alpha_j$ | $p_j$ | $w_j$ | P | C | W |
|---|---|---|---|---|---|---|
| 1 | $2.33 \times 10^{-5}$ | 1 | 7 | | | |
| 2 | $1.45 \times 10^{-5}$ | 2 | 8 | 110 | 175 | 200 |
| 3 | $5.41 \times 10^{-6}$ | 3 | 8 | | | |
| 4 | $8.05 \times 10^{-5}$ | 4 | 6 | | | |
| 5 | $1.95 \times 10^{-5}$ | 2 | 9 | | | |

$$\beta_j = 1.5, \quad j = 1,2,3,4,5, \quad t = 1000$$

Table 2  Optimal Solution for Example 13.1.

| | $R_1$ | $R_2$ | $R_3$ | $R_4$ | $R_5$ |
|---|---|---|---|---|---|
| Starting point | 0.70 | 0.70 | 0.70 | 0.70 | 0.70 |
| Initial step size | 0.05 | | | | |
| Final step size | 0.00039 | | | | |

| | $R_1$ | $R_2$ | $R_3$ | $R_4$ | $R_5$ | $X_1$ | $X_2$ | $X_3$ | $X_4$ | $X_5$ |
|---|---|---|---|---|---|---|---|---|---|---|
| Optimal point | 0.7582 | 0.8000 | 0.9000 | 0.8000 | 0.7500 | 3 | 3 | 2 | 2 | 3 |

Optimal system reliability 0.91494

Slack for the first constraint = 28

Slack for the second constraint = 0.033727

Slack for the third constraint = 1.4118

Table 3    Optimal Solution for Example 13.2.

| | $R_1$ | $R_2$ | $R_3$ | $R_4$ | $R_5$ | $X_1$ | $X_2$ | $X_3$ | $X_4$ | $X_5$ |
|---|---|---|---|---|---|---|---|---|---|---|
| Starting point | 0.7 | 0.7 | 0.7 | 0.7 | 0.7 | | | | | |
| Optimal point | 0.900 | 0.850 | 0.856 | 0.750 | 0.850 | 3 | 4 | 4 | 4 | 4 |

Optimal system reliability, $R_5$ = 0.993657

Slack for the first constraint = 35

Slack for the second constraint = 0.033247

Slack for the third constraint  = 18.476

Initial step size = 0.05

Final step size = 0.0002

Table 4    An Alternate Optimal Solution for Example 13.2.

| | $R_1$ | $R_2$ | $R_3$ | $R_4$ | $R_5$ | $X_1$ | $X_2$ | $X_3$ | $X_4$ | $X_5$ |
|---|---|---|---|---|---|---|---|---|---|---|
| Starting point | 0.8 | 0.8 | 0.8 | 0.8 | 0.8 | | | | | |
| Optimal point | 0.850 | 0.863 | 0.902 | 0.700 | 0.900 | 4 | 4 | 3 | 5 | 3 |

Optimal system reliability, $R_s$ = 0.994767

Slack for the first constraint = 27

Slack for the second constraint = 0.006542

Slack for the third constraint  = 24.226

Initial step size = 0.05

Final step size   = 0.0002

value of the systems reliability at the optimum is quite flat; therefore, there is a flexibility to select various values of component
reliabilities and redundancies which have nearly the same optimal system reliability.

## 13.5   CONCLUDING REMARKS

The determination of the optimal number of redundancies as well
as the optimal component reliability level in each of stages are carried out by a combination of the well-known Hooke and Jeeves pattern
search technique and a heuristic approach.  The optimal system reliability problem is an extension of the usual reliability optimization
problem and is a mixed integer nonlinear programming problem.  The
heuristic approach insures the integer number of redundancies with
nonlinear constraints, while the Hooke and Jeeves pattern search optimizes the component reliability level.  This procedure seems to be
very efficient in solving this problem.

## REFERENCES


1.  Aggarwal, K. K., J. S. Gupta, and K. B. Misra, "A New heuristic
    criterion for solving a redundancy optimization problem," *IEEE
    Transactions on Reliability*, Vol. R-24, No. 1, pp. 86-87 (1975).

2.  Hooke, R., and T. A. Jeeves, "Direct search solutions of numer-
    ical and statistical problems," *J. Assoc. Compt. Mach.*, Vol. 8,
    pp. 212-224 (1961).

3.  Misra, K. B., and M. D. Ljubojevic, "Optimal reliability design
    of a system:  a new look," *IEEE Transactions on Reliability*,
    Vol. R-22, pp. 255-258 (1973).

4.  Sharma, J., and K. V. Venkateswaran, "A direct method for maxi-
    mizing the system reliability," *IEEE Transactions on Reliability*,
    Vol. R-20, No. 4, pp. 256-259 (1971).

5.  Tillman, F. A., and J. M. Littschwager, "Integer programming
    formulation of constrained reliability problems," *Management
    Science*, Vol. 13, No. 11, pp. 887-899 (1967).

6.  Tillman, F. A., C. L. Hwang, L. T. Fan, and S. A. Balbale, "Systems reliability subject to multiple nonlinear constraints," *IEEE Transactions on Reliability,* Vol. R-17, No. 3, pp. 153-157 (1968).

7.  Tillman, F. A., C. L. Hwang, and W. Kuo, "Optimization techniques for systems reliability with redundancy - a review," *IEEE Transactions on Reliability,* Vol. R-26, pp. 148-155 (1977).

Dynamic programming provides a powerful tool for solving multistage
decision processes which arise in a number of fields.  It is based
on the so-called "principle of optimality" and employs the techniques
of invariant imbedding.  The essential notion of dynamic programming
is that problems are linked to a serial structure.  As mentioned its
cornerstone is the principle of optimality developed by Bellman in
 1957  [1] which states, "An optimal policy has the property that what-
ever the initial state and initial decisions are, the remaining decisions must constitute an optimal policy with regard to the state resulting from the first decision."

Consider a multistage process for which $x_n$ denotes a state vector
which represents a set of variables from stage n, and $\theta_n$ is a decision
(or control) vector which stands for a set of decision (or control)
variables at stage n.

The notion of a stage is actually an abstract one and the function of each stage is to transform the state variables from the input
state to the output state.  This transformation can generally be expressed as

$$x_n = T_n(x_{n+1};\ \theta_n), \qquad n = N, N-1, \ldots, 2, 1 \tag{1}$$

Eq. (1) is of a vector form.  If there are s state variables and one
decision variable, eq. (1) can be written as

$$x_{i,n} = T_{i,n}(x_{1,n+1}, \ x_{2,n+1}, \ \ldots, \ x_{s,n+1}, \ \theta_n) \tag{2}$$

The objective for the optimization of a multistage process is
to seek a set of admissible values of $\theta_1, \ \theta_2, \ \ldots, \ \theta_N$ so that a desired performance criterion or a return function is maximized (or
minimized). An inherent feature of multistage decision processes is
that there is an interval profit or return associated with each stage
of the process. The objective function is expressed as the summation
of these interval profits,

$$S(x_{N+1}; \ \theta_N, \ \ldots, \ \theta_2, \ \theta_1) = \sum_{n=N}^{1} g_n(x_{n+1}; \ \theta_n) \tag{3}$$

The value of the objective function depends upon the initial state
and the sequence of the decisions, $\theta_N, \ \ldots, \ \theta_2, \ \theta_1$. If we represent
the maximum return or objective function by $f_N(x_{N+1})$, then

$$f_N(x_{N+1}) = f_N(x_{1,N+1}, \ x_{2,N+1}, \ \ldots, \ x_{s,N+1})$$

$$= \max \ S(x_{N+1}; \ \theta_N, \ \ldots, \ \theta_1)$$

$$= \max_{\{\theta_n\}} \ \sum_{n=N}^{1} g_n(x_{n+1}; \ \theta_n) \tag{4}$$

Thus, in general, $f_n(x_{n+1})$ is the maximum return obtainable from the
operation of an n-stage process if an optimal policy is followed starting with the initial state, $x_{n+1}$.

   If there is one decision variable at each stage, eq. (4) becomes
an N-dimensional optimization problem because in that it must be optimized with respect to all N decision variables.  The dynamic programming technique treats with this problem as N one-dimensional problems.  For a one-stage process, eq. (4) becomes

$$f_1(x_2) = \max_{\theta_1} \{g_1(x_2; \theta_1)\} \tag{5}$$

which is the simplest optimization problem among the sequence of problems for $n = 1, 2, \ldots, N$. The other members of this sequence can be obtained by writing eq. (4) in the form,

$$f_n(x_{n+1}) = \max_{\theta_n} \max_{\theta_{n-1}} \ldots \max_{\theta_1} \{g_n(x_{n+1}; \theta_n) + \ldots + g_1(x_2; \theta_1)\}$$

Since the inputs to the stages following stage n are all affected by $\theta_n$ and the state of stage n is not affected by decisions made at stages following it, we can rewrite this as

$$f_n(x_{n+1}) = \max_{\theta_n} \left\{ g_n(x_{n+1}; \theta_n) + \max_{\theta_{n-1}} \ldots \max_{\theta_1} \left[ g_{n-1}(x_n; \theta_{n-1}) \right.\right.$$

$$\left.\left. + \ldots + g_1(x_2; \theta_1) \right] \right\} \tag{6}$$

The expression

$$\max_{\theta_{n-1}} \ldots \max_{\theta_1} \left( g_{n-1}(x_n; \theta_{n-1}) + \ldots + g_1(x_2; \theta_1) \right)$$

stands for the maximum return (or objective function) from an $(n-1)$ stage process with an initial state $x_n$. Hence, we can write

$$f_{n-1}(x_n) = \max_{\theta_{n-1}} \ldots \max_{\theta_1} \left( g_{n-1}(x_n; \theta_{n-1}) + \ldots + g_1(x_2; \theta_1) \right) \tag{7}$$

Eq. (6) can be further simplified to

$$f_n(x_{n+1}) = \max_{\theta_n} \left( g_n(x_{n+1}; \theta_n) + f_{n-1}(x_n) \right)$$

or

$$f_n(x_{n+1}) = \max_{\theta_n} \left( g_n(x_{n+1}; \theta_n) + f_{n-1}(T(x_{n+1}; \theta_n)) \right) \qquad (8)$$

This last expression is the so-called functional equation and is a mathematical statement of the principle of optimality.  It states the recursive relationship between an n stage process and an n-1 stage process.  The solution yields the value for the maximum return with the corresponding optimal policy as a function of the set $\{\theta_n\}$.

Further details dealing with dynamic programming as an optimization tool are in texts by Bellman [1] and Bellman and Dreyfus [2].


## REFERENCES


1.  Bellman, R., *Dynamic Programming*, Princeton, N.J.: Princeton University Press (1957).

2.  Bellman, R., and S. E. Dreyfus, *Applied Dynamic Programming*, Princeton, N. J.: Princeton University Press (1962).

Consider a simple multistage process consisting of N stages connected in series.  The state of the process stream denoted by an s-dimensional vector, $\bar{x} = (x_1, x_2, \ldots, x_s)$, is transformed at each stage according to an r-dimensional decision vector, $\bar{\theta} = (\theta_1, \theta_2, \ldots, \theta_r)$, which represents the decisions made at that stage.  The transformation of the process stream at the $n^{th}$ stage is described by a set of performance equations in vector form.

$$\bar{x}^n = \bar{T}^n(\bar{x}^{n-1}; \bar{\theta}^n), \qquad n = 1,2,\ldots,N \qquad (1)$$

$$\bar{x}^0 = \bar{\alpha}$$

where the superscript n indicates the stage number.  The exponents are written with parentheses or brackets such as $(x^n)^2$ or $[T^n(x^{n-1}; \theta^n)]^2$.

A typical optimization problem associated with such a process is to find a sequence of $\bar{\theta}^n$, n = 1,2,...,N subject to the constraints

$$\psi_i^n [\theta_1^n, \theta_2^n, \ldots, \theta_r^n] \leq 0 \qquad (2)$$

$$n = 1,2,\ldots,N; \qquad i = 1,2,\ldots,r$$

which makes a function of the state variable of the final stage

$$S = \sum_{i=1}^{s} c_i x_i^N, \qquad c_i = \text{constant} \qquad (3)$$

an extremum when the initial condition $\bar{x}^0 = \bar{\alpha}$ is given.  The function, S, which is to be maximized (or minimized), is the objective function of the process.

The procedure for solving such an optimization problem by the discrete maximum principle is to introduce an s-dimensional adjoint vector $\bar{z}^n$ and a Hamiltonian function $H^n$, which satisfy the following relations:

$$H^n = (\bar{z}^n)^T \bar{x}^n = \sum_{i=1}^{s} z_i^n T_i^n (\bar{x}^{n-1}; \bar{\theta}^n), \qquad n = 1,2,\ldots,N \qquad (4)$$

$$z_i^{n-1} = \frac{\partial H^n}{\partial x_i^{n-1}}, \qquad i = 1,2,\ldots,s; \qquad n = 1,2,\ldots,N \qquad (5)$$

and

$$z_i^N = c_i, \qquad i = 1,2,\ldots,s \qquad (6)$$

If the optimal decision vector function $\theta*^n$, which makes the objective function S an extremum, is interior to the set of admissible decisions $\bar{\theta}^n$, the set given by eq. (2); a necessary condition for S to be an extremum (local) with respect to $\bar{\theta}^n$, is

$$\frac{\partial H^n}{\partial \bar{\theta}^n} = 0, \qquad n = 1,2,\ldots,N \qquad (7)$$

If $\theta*^n$ is at a boundary of the set, it can be determined from the condition that $H^n$ is an extremum (locally).

Further details on the discrete maximum principle can be found in the text by Fan and Wang [1] and [2].

## REFERENCES


1. Fan, L. T., and C. S. Wang, *The Discrete Maximum Principle,* New York:  Wiley (1964).

2. Hwang, C. L., and L. T. Fan, "A discrete version of Pontryagin's maximum principle," *Operations Research,* Vol. 15, No. 1, pp. 139-146 (1967).

APPENDIX A3   OUTLINE OF THE GENERALIZED REDUCED GRADIENT METHOD (GRG)

The generalized reduced gradient method is a nonlinear programming
method proposed by Abadie and Carpentier [1, 2, 3].  The method is
essentially a generalization of the Wolfe reduced gradient technique
[6], which solves problems having a nonlinear objective function and
linear (equality) constraints.  In the Wolfe method, the variables
are classed as independent and dependent.  From the set of linear
(equality) constraints, the dependent variables are obtained in terms
of the independent variables, and the expressions thus obtained are
substituted into the objective function.  The original problem, therefore, is reduced to an unconstrained one with reduced dimension.
A variety of optimization techniques may then be used to find the
optimum solution.  Applying the same concepts to problems with nonlinear constraints adds to the computational difficulties, but is
not altogether impossible.

The general nonlinear programming problem with nonlinear equality constraints is defined as follows:

Determine vector $\bar{X}$ so as to maximize

$$f_0(\bar{X}) \tag{1}$$

subject to the constraints:

$$f(\bar{X}) = 0 \tag{2}$$

and the boundary conditions:

$$\bar{a} \le \bar{X} \le \bar{b} \tag{3}$$

where $\bar{X}$, $\bar{a}$, and $\bar{b}$ are N-dimensional column vectors, and $\bar{f}(\bar{X})$ is an M-dimensional column vector of constraint functions in terms of vector $\bar{X}(N > M)$. Inequality constraints may be employed by the appropriate addition of slack variables.

The problem is solved by partitioning the vector of variables into the independent and dependent sets of variables, of N-M and M dimensions, respectively. Let

$$\bar{X} = [\bar{x}, \bar{y}] \tag{4}$$

where $\bar{x}$ is the (N-M)-dimensional set of independent (basic) variables, and $\bar{y}$ be the M-dimensional set of dependent (nonbasic) variables. If the constraint functions satisfy the requirements of the Implicit Function Theorem [4], then the non-degeneracy assumption is satisfied that the dependent variables can be expressed as functions of the independent variables, i.e.,

$$\bar{y} = \bar{\phi}(\bar{x}) \tag{5}$$

where $\bar{y}$ is within the boundary:

$$\bar{a} \le \bar{X} = [\bar{x}, \bar{y}] \le \bar{b} \tag{6}$$

When this condition does not hold, the basis is changed until a feasible solution is obtained.

By substituting the vector $\bar{y}$ into the objective function, the problem may now be simply defined as:

$$\text{Maximize} \qquad f_0(\bar{X}) = f_0(\bar{x}, \bar{y}) = f_0(\bar{x}, \bar{\phi}(\bar{x})) \equiv F(\bar{x}) \tag{7}$$

subject to

$$\bar{a} \le \bar{x} \le \bar{b} \tag{8}$$

*Computational Procedure.*  The procedure for using the GRG method is summarized below (see Fig. 1) [1, 5]:

Step 1.  Compute the direction of movement, $\bar{h}^o$, at the starting point $\bar{X}^o = [\bar{x}^o, \bar{y}^o]$, by computing the "reduced gradients" at this point:

$$\bar{g}^{oT} = \frac{\partial F(\bar{x})}{\partial \bar{x}^o} = \frac{\partial f}{\partial \bar{x}^o}_o + \frac{\partial f}{\partial \bar{y}^o}_o \cdot \frac{\partial \bar{y}^o}{\partial \bar{x}^o} \tag{9}$$

From eq. (2) we have:

$$\frac{\partial \bar{f}}{\partial \bar{x}} + \frac{\partial \bar{f}}{\partial \bar{y}} \cdot \frac{\partial \bar{y}}{\partial \bar{x}} = 0 \tag{10}$$

Solving for $\partial \bar{y}/\partial \bar{x}$, we obtain

$$\frac{\partial \bar{y}}{\partial \bar{x}} = - \left( \frac{\partial \bar{f}}{\partial \bar{y}} \right)^{-1} \cdot \frac{\partial \bar{f}}{\partial \bar{x}} \tag{11}$$

Substituting eq. (11) into eq. (9) gives:

$$\bar{g}^{oT} = \frac{\partial f}{\partial \bar{x}^o}_o - \frac{\partial f}{\partial \bar{y}^o}_o \left( \frac{\partial \bar{f}}{\partial \bar{y}^o} \right)^{-1} \cdot \frac{\partial \bar{f}}{\partial \bar{x}^o} \tag{12}$$

The "projected reduced gradient", $\bar{p}^o$, for each component of the independent vector $\bar{x}$, is computed in the following manner:

$$p_i^o = \begin{cases} 0 & \text{if } x_i = \text{lower bound and } g_i^o \le 0 \\ 0 & \text{if } x_i = \text{upper bound and } g_i^o \ge 0 \\ g_i^o & \text{otherwise} \end{cases} \tag{13}$$

$$i = 1, 2, \ldots, N-M$$

Now

$$\bar{h}^o = \bar{p}^o \tag{14}$$

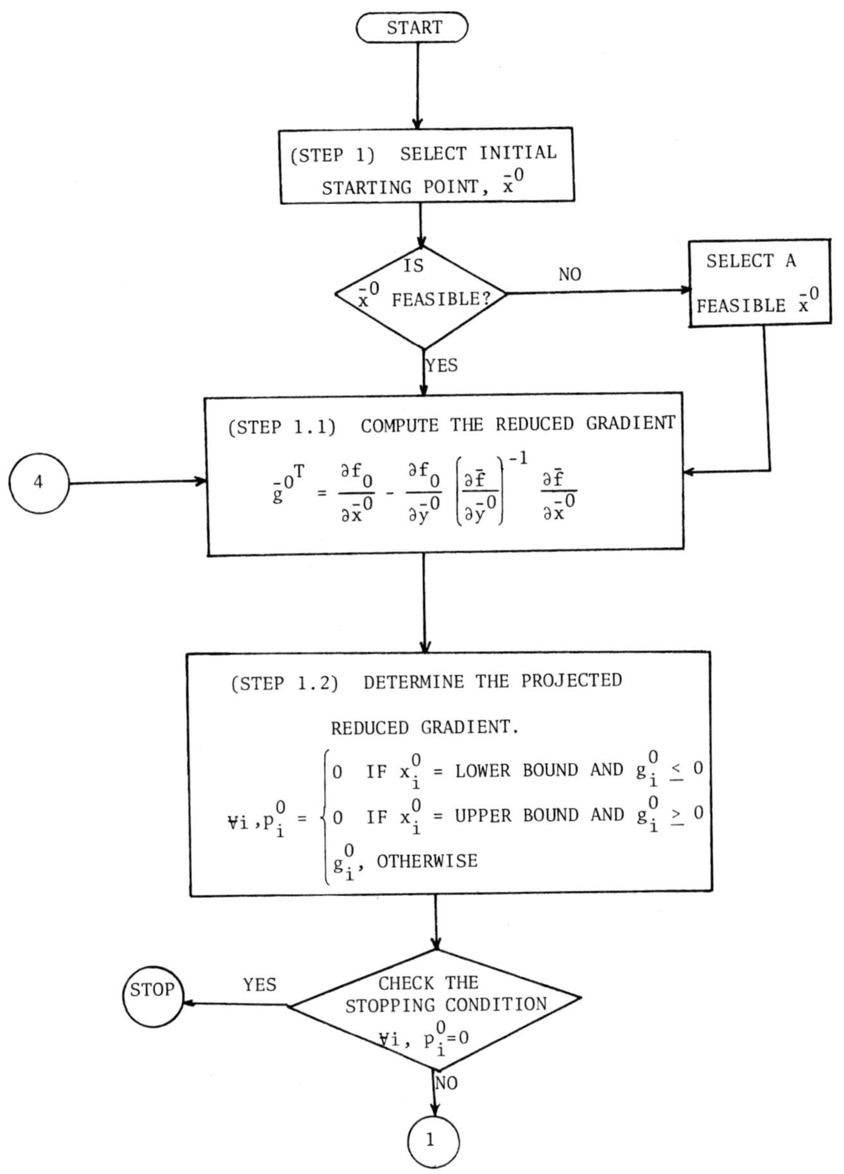



Figure 1   Computer Flow Diagram for the GRG Algorithm.

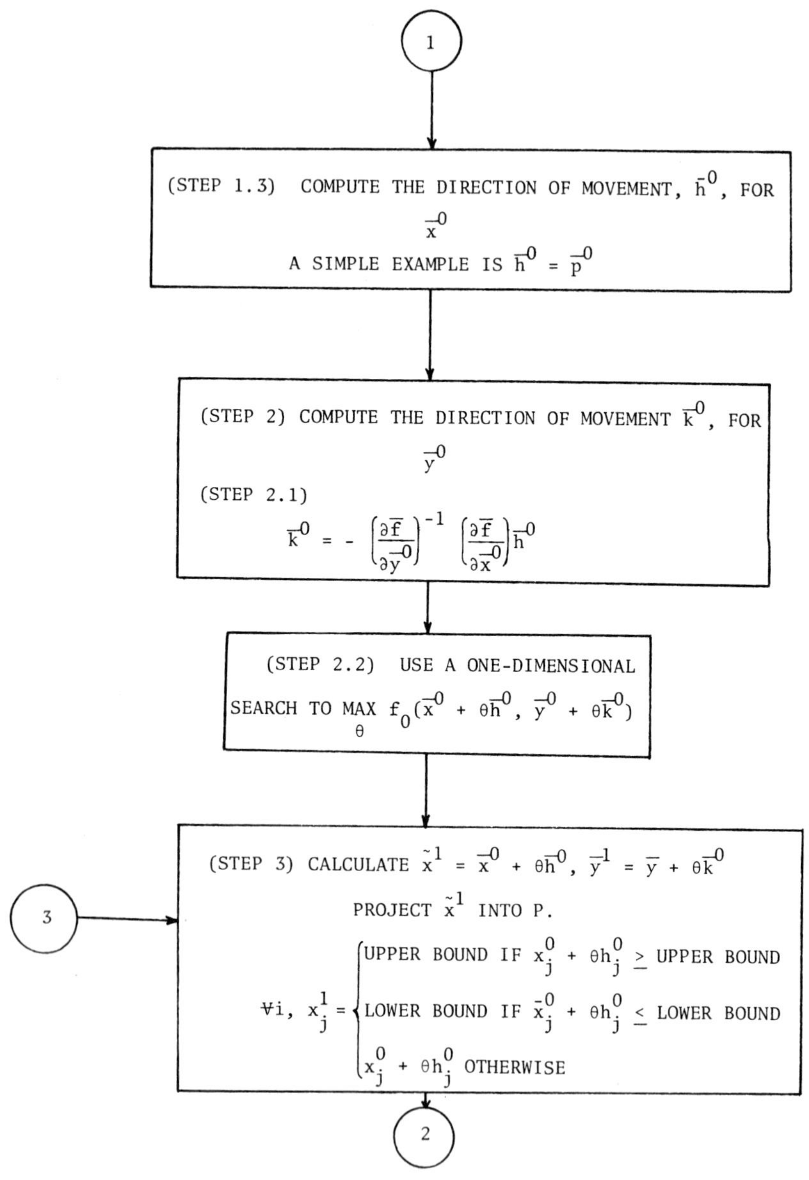


Figure 1   (Continued)

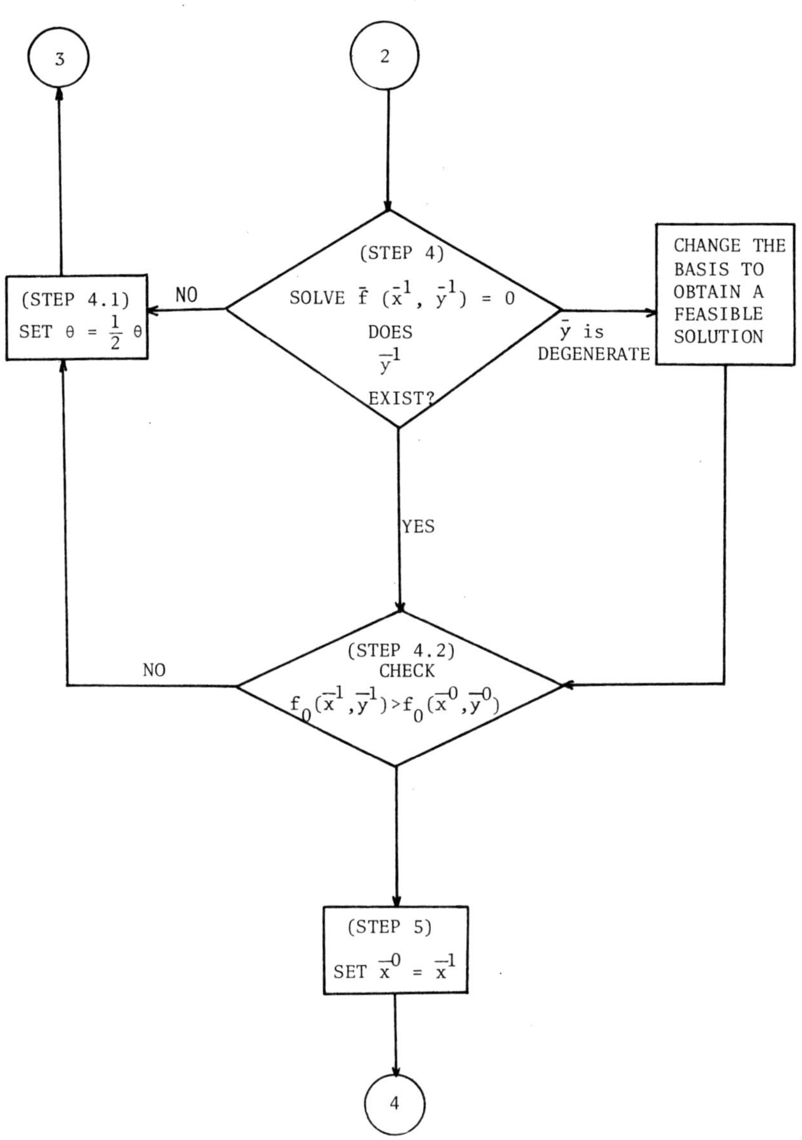


Figure 1   (Continued)

Step 2.  It is desirable to stay in the feasible region, or at least close to it, by selecting a proper direction of movement.  The steps for vectors $\bar{x}^o$ and $\bar{y}^o$ are $\bar{x}^o + \theta\bar{h}^o$, and $\bar{y}^o + \theta\bar{K}^o$, respectively.  The desired movement is along the surface of the constraints.  This is accomplished by finding the tangent to $\bar{f}(\bar{x}^o + \theta\bar{h}^o, \bar{y}^o + \theta\bar{K}^o) = 0$ at the point $(\bar{x}^o, \bar{y}^o)$, that is:

$$\frac{\partial \bar{f}}{\partial \bar{x}^o} \cdot \bar{h}^o + \frac{\partial \bar{f}}{\partial \bar{y}^o} \cdot \bar{K}^o = 0 \tag{15}$$

This yields

$$\bar{K}^o = - \left( \frac{\partial \bar{f}}{\partial \bar{y}^o} \right)^{-1} \left( \frac{\partial \bar{f}}{\partial \bar{x}^o} \right) \bar{h}^o \tag{16}$$

where

$$\bar{f}(\bar{x}^o + \theta\bar{h}^o, \bar{y}^o + \theta\bar{K}^o)$$

is to be optimized for $\theta$ using a one-dimensional search technique.

Step 3.  After calculating:

$$\tilde{x}^1 = \bar{x}^o + \theta\bar{h}^o$$

$$\tilde{y}^1 = \bar{y}^o + \theta\bar{K}^o$$

$$f_o(\tilde{x}^1, \tilde{y}^1),$$

the values of the independent variables are projected  within the bounds

$a \leq X \leq b$ as follows:

$$x_j^1 = \begin{cases} \text{lower bound if } x_j^o + \theta h_j^o \leq \text{lower bound} \\ \text{upper bound if } x_j^o + \theta h_j^o \geq \text{upper bound} \\ x_j^o + \theta h_j^o \text{ otherwise} \\ j = 1,2, \ldots, N\text{-}M \end{cases} \tag{17}$$

Step 4.  A feasible solution is determined by solving the following expression iteratively by:

$$f(\bar{x}^{-1}, \bar{y}^{-1}) = 0$$

The existence of $\bar{y}^{-1} = \bar{\phi}(\bar{x}^{-1})$ is insured by the Implicit Function Theorem as mentioned before.  If a component of $\bar{y}^{-1}$ violates a boundary condition (degeneracy), a change of basis occurs.  Two cases may arise at the end:

a)  if the iterative procedure does not converge to a $\bar{y}^{-1}$, then $\bar{x}^{-1}$ is out of the functional domain.  This is alleviated by reducing $\theta$ and returning to Step 3.

b)  if the solution obtained is $\bar{y}^{-1}$, then the solution vector is $\bar{x}^{-1} = [\bar{x}^{-1}, \bar{y}^{-1}]$.  If the solution vector does not improve the objective function, $\theta$ is reduced by half and the procedure starts again at Step 3.

Step 5.  $\bar{x}^{o}$ is set equal to $\bar{x}^{1}$ and the algorithm is repeated.  However, if a better value for $\theta$ can be somehow determined, a return to Step 3 is made before the iteration proceeds.

The termination criterion for the GRG method is, theoretically, when:

$$\bar{p}^{o}_{i} = 0, \qquad i = 1,2, \ldots, N-M$$

In practice, however, the following three stopping criteria are used:

$$1) \quad ||\bar{p}^{o}|| = \sum_{i=1}^{N-M} (p^{o}_{i})^{2} < \varepsilon_{1}$$

$$2) \quad p^{o}_{i} < \varepsilon_{2} \tag{18}$$

$$3) \quad |f_{o}(\bar{x}^{1}) - f_{o}(\bar{x}^{o})| < \varepsilon_{3}$$

The GRG method has been studied extensively, and coded in FORTRAN by Abadie and Guigou [1, 2, 3].  Three generations of the program have been developed.  The first, called GRG 66, was an experimental code, followed by the second one, GRG 69.  An improved code,

GREG, is the outgrowth of the first two and by far the best of the three.  It can be obtained by writing J. Abadie, Electricite de France, Paris, France.


## REFERENCES


1.  Abadie J., "Application of the GRG algorithm to optimal control problems,"in *Integer and Nonlinear Programming,* J. Abadie, (ed:) Amsterdam: North Holland Publishing Co. (1970).

2.  Abadie, J. and J. Carpentier, "Generalization of the Wolfe reduced gradient method to the case of nonlinear constraints," in *Optimization,* R. Fletcher, (ed.), New York: Academic Press (1969).

3.  Abadie, J. and J. Guigou, "Numerical experiments with the GRG method," in *Integer and Nonlinear Programming,* J. Abadie, (ed.), Amsterdam: North Holland Publishing Co. (1970).

4.  Apostol, T. M., *Mathematical Analysis,* Reading, Mass: Addison-Wesley (1957).

5.  Hwang, C. L., J. L. Williams, and L. T. Fan, "Introduction to the generalized reduced gradient method," Institute for Systems Design and Optimization, Report No. 39, Kansas State University, Manhattan, Kansas (1972).

6.  Wolfe, P., "Methods of nonlinear programming," in *Recent Advances in Mathematical Programming,* R. Graves and P. Wolfe, (eds.) New York:  McGraw-Hill (1963).

By employing the inequality which states that the arithmetic mean
is at least as great as the geometric mean, a dual problem for many
optimal design problems may be formulated. Geometric programming uses
this inequality and the relationships of the primal and dual problems
to solve many optimization problems. The primal problem is expressed
in terms of a class of functions which we call positive polynomials,
or posynomials for short.

The primal problem is that of minimizing a posynomial S subject
to constraints of a certain type.  Let M denote the constrained minimum value of the primal function S.  Because of the inequality relating
the arithmetic and geometric means, there is a related maximization
problem concerning a function v which is the dual function.  It will
be shown that the dual problem is one of maximizing v subject to certain linear constraints.  We will also show that M is the constrained
maximum value of v as well as the constrained minimum value of S.

Geometric programming is based on the concept that the arithmetic mean is greater than or equal to the geometric mean inequality.
For the general case, the weighted arithmetic and geometric means
satisfy the relation

$$\sum_{i=1}^{n} \delta_i U_i \geq \prod_{i=1}^{n} U_i^{\delta_i} \tag{1}$$

where the $\delta_i$'s are the weights which must sum to unity, which states
that the normality condition,

$$\delta_1 + \delta_2 + \ldots + \delta_n = 1 \tag{2}$$

must be satisfied.  Eq. (1) is an equality if and only if all of the
$U_i$'s are equal.

Now, suppose we wish to find the minimum value of the objective
function

$$S = \frac{x_2}{x_1^2} + x_2 + 2 \frac{x_1}{x_2} \tag{3}$$

we have

$$S = \frac{1}{4}\left(\frac{4x_2}{x_1^2}\right) + \frac{1}{4}(4x_2) + \frac{2}{4}\left(\frac{4x_1}{x_2}\right) \geq \left(\frac{4x_2}{x_1^2}\right)^{\frac{1}{4}} (4x_2)^{\frac{1}{4}} \left(\frac{4x_1}{x_2}\right)^{\frac{2}{4}} = 4 \tag{4}$$

From this equation we find that 4 is a lower bound for S, that is,

$$S \geq 4 \tag{5}$$

Using differential calculus, we can show that 4 is the minimum value
of S and that this occurs at $x_1 = x_2 = 1$.

The preceding example illustrates how we can obtain the minimum
value of an objective function directly by properly choosing the
weights of each term in the posynomial.  If the geometric mean is
properly weighted, it is independent of its variables, and it is not
necessary to determine the values of the variables prior to finding
the minimum value of the objective function.  It is this unique property of the geometric mean that makes it easy to minimize certain
posynomials.

The weighted arithmetic mean - geometric mean inequality with
the normality condition can be written as

$$u_1 + u_2 + \ldots + u_n \geq \left(\frac{u_1}{\delta_1}\right)^{\delta_1} \left(\frac{u_2}{\delta_2}\right)^{\delta_2} \cdots \left(\frac{u_n}{\delta_n}\right)^{\delta_n} \tag{6}$$

if we let $u_i = \delta_i U_i$ for $i = 1,2,\ldots,n$.

The left side of this inequality is the posynomial S that is to be minimized. For simplicity, we shall refer to eq. (6) as the geometric inequality, and we shall call the left side the primal function and the right side the predual function. Using V to denote the predual function, the inequality, eq. (6), becomes

$$S \geq V \tag{7}$$

If the primal function is a posynomial and the $u_j$ is given by

$$u_j = C_j \prod_{i=1}^{m} x_i^{a_{ji}}, \qquad C_j > 0 \tag{8}$$

Substituting the above into

$$V = \left(\frac{u_1}{\delta_1}\right)^{\delta_1} \left(\frac{u_2}{\delta_2}\right)^{\delta_2} \cdots \left(\frac{u_n}{\delta_n}\right)^{\delta_n} \tag{9}$$

gives

$$V = \left(\frac{C_1}{\delta_1}\right)^{\delta_1} \left(\frac{C_2}{\delta_2}\right)^{\delta_2} \cdots \left(\frac{C_n}{\delta_n}\right)^{\delta_n} x_1^{D_1} x_2^{D_2} \cdots x_m^{D_m} \tag{10}$$

where

$$D_j = \sum_{i=1}^{n} \delta_i a_{ij}, \qquad j = 1,2,\ldots,m \tag{11}$$

Sometimes it is possible to choose the weights $\delta_i$ in such a way that all of the exponents $D_j$ are zero. When this is possible, the predual function, V, does not depend upon the variables $x_1, x_2, \ldots, x_n$. When all of the $D_j$ are zero, eq. (10) becomes

$$v = \left(\frac{C_1}{\delta_1}\right)^{\delta_1} \left(\frac{C_2}{\delta_2}\right)^{\delta_2} \cdots \left(\frac{C_n}{\delta_n}\right)^{\delta_n} \tag{12}$$

which we shall refer to as the dual function, v.

From inequality (7), we know that our objective function, S, has
a minimum point.  We shall use M to designate this positive greatest
lower bound of S which must satisfy the inequality

$$S \geq M \geq v \tag{13}$$

We shall now consider the more general primal problem of minimizing a posynomial subject to p inequality constraints and the corresponding dual problem of maximizing the dual function subject to
its constraints. A complete statement of the primal problem is as
follows:

*Primal Problem.*  Find the minimum value of an objective function

$$S = \sum_{i=1}^{n_0} u_i \tag{14}$$

subject to the constraints

$$g_1 \leq 1, \; g_2 \leq 1, \; \ldots, \; g_p \leq 1 \tag{15}$$

$$g_k = \sum_{i=m_k}^{n_k} u_i, \qquad k = 1,2,\ldots,p \tag{16}$$

Here

$$m_k = n_{k-1} + 1, \qquad k = 1,2,\ldots,p \tag{17}$$

and the $u_i$ in eqs. (14) and (16) are numbered consecutively from 1
to $n_p = n$.  The $u_i$ are defined as follows:

$$u_i = C_i \, x_1^{a_{i1}} \, x_2^{a_{i2}} \, \ldots \, x_m^{a_{im}}, \qquad i = 1,2,\ldots,n \tag{18}$$

where

$$x_1 > 0, \; x_2 > 0, \; \ldots, \; x_m > 0 \tag{19}$$

The exponents $a_{ij}$ are arbitrary real numbers, but the coefficients $C_i$ are assumed to be postive, that is, the objective function S and the constraint functions $g_k$ are posynomials.  The posynomial S, which is to be minimized, is a function of m independent variables $x_1, x_2, \ldots, x_m$.

The dual problem, which corresponds to the primal problem, is as follows:

*Dual Problem.*  Find the maximum value of the dual function

$$v = \left\{ \prod_{i=1}^{n} \left( \frac{C_i}{\delta_i} \right)^{\delta_i} \right\} \prod_{k=1}^{p} \lambda_k^{\lambda_k} \tag{20}$$

where

$$\lambda_k = \sum_{i=m_k}^{n_k} \delta_i, \qquad k = 1,2,\ldots,p \tag{21}$$

Here

$$m_1 = n_o + 1, \; m_2 = n_1 + 1, \; \ldots, \; m_p = n_{p-1} + 1$$

The constants $C_i$ are assumed to be positive and the weights $\delta_1, \delta_2, \ldots, \delta_n$ are subject to the linear constraints

$$\delta_1 \geq 0, \; \delta_2 \geq 0, \; \ldots, \; \delta_n \geq 0 \tag{22}$$

$$\sum_{i=1}^{n_o} \delta_i = 1 \tag{23}$$

$$\sum_{i=1}^{n} a_{ij} \delta_i = 0, \qquad j = 1,2,\ldots,m \tag{24}$$

where the coefficients, $a_{ij}$, are all real numbers.

The dual function, v, is a function of the variables, $\delta_1, \delta_2, \ldots, \delta_n$, and the linear constraints of the positivity condition (eq. (22)), the normality condition (eq.(23)), and the orthogonality condition (eq. (24)) which are all imposed on these variables.

Note the manner in which the dual problem is generated from its
corresponding primal problem.  The positive constants, $C_i$, appearing
in the dual function, $v$, are the coefficients of the posynomials whose
terms are given by eq. (18).  Each $\delta_i$ is associated with the $i^{th}$
term, $u_i$, of the primal problem, and hence, each $u_i$ of the posynomials
is associated with one and only one of the dual variables, $\delta_1$, $\delta_2$,
..., $\delta_n$.  Each $\lambda_k$ in the dual problem comes from a forced constraint,
$g_k \leq 1$, of the primal problem.

Further details on how to solve optimization problems with constraints using geometric programming are provided in the texts by
Duffin et al. [1], Wilde and Beightler [2], and Zener [3].

## REFERENCES


1.  Duffin, R. J., E. L. Peterson, and C. Zener, *Geometric Programming*,
    New York: Wiley (1966).

2.  Wilde, D. J. and C. S. Beightler, *Optimization Theory*, Englewood
    Cliffs, N. J.: Prentice-Hall (1967).

3.  Zener, C., *Engineering Design by Geometric Programming*, New York:
    Wiley-Interscience (1971).